**ALSO BY
THOMAS LEVENSON**

*Newton and the Counterfeiter: The Unknown Detective Career
of the World's Greatest Scientist*

Einstein in Berlin

Measure for Measure: A Musical History of Science

THE HUNT FOR

VULCAN

How Albert Einstein Destroyed a Planet

and Deciphered the Universe

THOMAS LEVENSON

HEAD
of ZEUS

Book design by Simon M. Sullivan

Head of Zeus Ltd
Clerkenwell House
45–47 Clerkenwell Green
London EC1 R 0HT

WWW.HEADOFZEUS.COM

For Katha and Henry, ever-renewing gifts,
and in remembrance of my uncles
Daniel Levenson and David Sebag-Montefiore,
who stood up through the hard years
and the fine ones

Contents

Preface

November 18, 1915, Berlin.

A man is on the move, coming into the center of town from the western suburbs. Usually a bit disheveled—his shock of hair would become almost independently famous—today he's fully presentable, girded for public performance. He enters Unter den Linden, the grand avenue that pierces the Brandenburg Gate on the way east to the River Spree. He walks up to number 8, the entrance to the Prussian Academy of Sciences, and steps inside.

On this Thursday in the second autumn of what was already being called the "Great War," the members of the Academy settle themselves in for a lecture, the third of four in a row by one of their newest colleagues. That still-young man makes his way to the front of the room. He takes up his notes—just a few pages—and begins to speak.

Albert Einstein's talk that day and its sequel, presented the following week, completed the greatest individual intellectual accomplishment of the twentieth century. We now call that idea the general theory of relativity: at once a theory of gravity and the foundation for the science of cosmology, the study of the birth and evolution of the universe as a whole. Einstein's results mark the triumph of a lone thinker, battling the odds, the doubts of his peers, and the most famous scientist in history, Sir Isaac Newton.

For all the grand sweep of his theory, though, when he spoke on the 18th, Einstein focused on something much more parochial: Mercury, the smallest planet then known, and—at an even finer grain of detail—a tiny, unexplained hitch in its orbit, a wobble, barely measurable, for which there was (until he spoke) no adequate explanation.

By 1915, Mercury's misbehavior had been recognized for over sixty years. Throughout that time, astronomers had gone to greater and greater lengths to come up with some explanation for this errant behavior within the conventional framework of Newton's centuries-old account of gravity—the crowning victory in the Scientific Revolution. The first and seemingly most obvious idea imagined a whole new planet hidden in the glare of the sun, which could provide enough of a gravitational tug to haul Mercury out of its "correct" orbit.

As a hypothesis, conjuring a planet out of an orbital glitch was perfectly reasonable. Indeed, there was precedent, and at first it seemed not just logical, but right. Almost as soon as Mercury's plight became public knowledge, amateur and professional astronomers alike spotted and identified an object lurking within the concealing glare of the sun. It would be seen again, over and over, more than a dozen times over two decades. Its own orbit would be calculated; its history recovered in old records of unexplained sightings; it would even receive a name.

There was only one problem:

The planet Vulcan was never there.

This book tells Vulcan's story: its ancestry, its birth, its odd, twilit journey in and out of the grasp of eager would-be discoverers, its time in purgatory, and finally, on the 18th of November, 1915, its decisive end at the hands of Albert Einstein.

At first blush, this may seem something of a burlesque, a tale of nineteenth-century astronomical follies, Victorian gentlemen chasing a mistake. But there's more here than a comedy of errors.

The story of Vulcan suggests something much deeper, an insight that gets to the heart of the way science really advances (as opposed to the way we're taught in school).

The enterprise of making sense of the material world turns on a key question: what happens when something observed in nature doesn't fit within the established framework of existing human knowledge? The standard answer is that scientific ideas are supposed to evolve to accommodate new facts. After all, science is a uniquely powerful way of figuring things out precisely because all of its claims, even its most beloved, are subject to the ultimate test of reality. In our common description of the scientific method, any empirical result that refuses to conform to the demands of a theory invalidates that theory, and requires the construction of a new one.

Ideas, though, are hard to relinquish, none more so than those of Isaac Newton. For decades, the old understanding of gravity was so powerful that observers on multiple continents risked their retinas to gaze at the sun in search of Vulcan. And, contrary to the popular picture of science, a mere fact—Mercury's misplaced motion—wasn't nearly enough to undermine that sturdy edifice. As Vulcan's troublesome history reveals, no one gives up on a powerful, or a beautiful, or perhaps simply a familiar and useful conception of the world without utter compulsion—and a real alternative.

Einstein wrote Vulcan out of history on the third Thursday in the second November of the war. It had taken him the better part of a decade to develop what became his radical new picture of gravity: how matter and energy mold space and time; how space and time fix the paths that matter and energy must take. As presented to his colleagues that Thursday afternoon, Einstein showed how Mercury's "wobble" turned out to be just its natural path, the one it has to take in a universe in which relativity is true. That result emerged at the end of a chain of mathematical reasoning, the inevitable outcome of subjecting matter to number.

In that context, Vulcan's fate provided the first test of general relativity, proof that Einstein had managed to capture something true about how our universe works. But to get to that point, to follow the radical strangeness of general relativity all the way to its conclusion took both boldness and exquisitely subtle reasoning: hard labor sustained over the eight years it took Einstein to dispatch the ghost planet. That part of the story shows how powerful a thinker it took to clamber past accepted wisdom to achieve what he, alone of all his peers, was able to do.

Einstein, usually a fairly phlegmatic man, felt this one to the bone. When he completed the calculation of the orbit of Mercury and saw exactly the right number fall out of the long chain of pure reasoning, he told friends that he felt "beside himself with excitement." Seeing Mercury's motion simply fall out of his equations pierced him to his heart, he said. He felt palpitations, a sensation "as if something had burst within him."

Vulcan is long gone, almost completely forgotten. It may seem today to be merely a curiosity, just another mistake our ancestors made, about which we now know better. But the issue of what to do with failure in science was tricky right at the start of the Scientific Revolution, and it remains so now. We may—we do—know more than the folks back then. But we are not thus somehow immune to the habits of mind, the leaps of imagination, or the capacity for error that they possessed. Vulcan's biography is one of the human capacity to both discover and self-deceive. It offers a glimpse of how hard it is to make sense of the natural world, and how difficult it is for any of us to unlearn the things we think are so, but aren't.

And, in the end, it is a tale of the joy that accrues when we do.

Part One

NEWTON TO NEPTUNE

(1682–1846)

"THE IMMOVABLE ORDER OF THE WORLD"

August 1684, Cambridge.

Edmond Halley had suffered a sad and vexing spring. In March, his father disappeared under suspicious circumstances—a not-altogether-unusual fate in the political turmoil that shot through the last years of the Stuart dynasty's rule. He was found dead five weeks later. He'd left no will, which forced the younger Halley to spend the next few months dealing with the resulting mess: the twelve pounds owed to his father by a local rector; the three pounds a year promised as an annuity to a woman as part of a real estate transaction; rents to collect and trustees to satisfy. That miserable business consumed him into the summer, and ultimately required a trip to Cambridgeshire to handle face to face those details that couldn't be resolved from London.

There was nothing happy about the first part of that journey, but once he'd dealt with the legal issues, one unexpected pleasure came his way. In January, before his troubles began, Halley had produced a clever bit of celestial analysis, a calculation that suggested that whatever force held the planets on their paths around the sun grew weaker in proportion to the square of each object's distance from the sun. But that prompted an immediate question: could that particular mathematical relationship—called an inverse square law—explain why *all* celestial objects moved down the paths they'd been observed to follow?

The best minds in Europe knew what was at stake in that

seemingly technical issue. This was the decisive climax in what we've come to call the Scientific Revolution, the long struggle through which mathematics supplanted Latin as the language of science. On the 14th of January, 1684, following a meeting of the Royal Society, Halley fell into conversation with two old friends: the polymath Robert Hooke and the former president of the Society, Sir Christopher Wren. As their talk moved on to astronomy, Hooke claimed he'd already worked out the inverse square law that guided the motions of the universe. Wren didn't believe him, and so offered both Halley and Hooke a prize—a book worth roughly $300 in today's money—if either of them could present a rigorous account of such a universal law within two months. Halley swiftly acknowledged that he couldn't find his way to such a result, and Hooke, for all his bravado, failed to deliver a written proof by Wren's deadline.

There the matter stuck until, at last, Halley escaped from the wretchedness of postmortem wrangles with his surviving family. His business had taken him east from London anyway—why not detour to the university at Cambridge, there to gain at least an afternoon's respite in talk of natural philosophy? Coming into town he made his way to the great gate of the College of the Holy and Undivided Trinity. A left onto the college grounds, then right and almost immediately up the stairs would have brought him to the rooms occupied by the Lucasian Professor of Mathematics, Isaac Newton.

To most of his contemporaries, Newton in the summer of 1684 was something of an enigma. London's natural philosophers knew him as a man of formidable intelligence, but Halley was among very few who counted him as an acquaintance, much less a friend. The public record of Newton's work was slim.

His reputation rested on a handful of exceptional results, mostly transmitted to the secretary of the Royal Society in the early 1670s, but he was irascible, proud, swift to anger, and agonizingly slow to forgive, and an early dispute with Hooke left him unwilling to risk grubby public wrangling. He kept much of his work secret for the next decade—so much so that, as his biographer Richard Westfall put it, had he died in the spring of 1684, Newton would have been remembered as a very talented and rather odd man, and

Fashionable portraitist Godfrey Kneller painted the earliest known likeness of Isaac Newton in 1689.

nothing more. But those who made it so far as to be welcome in the rooms on the northeast corner of Trinity's Great Court would find someone capable of real warmth—and a mind whose power no learned man in Europe could match.

Much later Newton told the story of Halley's visit that summer day to another friend, and if the old man's memory wasn't playing tricks, the two men chatted about this and that for a while. But eventually Halley got down to the question troubling him since January: what about that inverse square relationship? What curve would the planets in their orbits trace, "supposing the force of the attraction towards the sun to be reciprocal to the square of their distance to it?"

Edmond Halley, painted by Thomas Murray around the time Principia *was published*

"An ellipse," Newton said instantly.

Halley, "struck with amazement and joy," asked how his friend knew that answer so surely.

"I have calculated it," Newton recalled telling his companion, and when Halley asked to see his workings, fumbled among his notes. On that day he claimed he couldn't find them, and promised to dig them up and send the result to Halley in London. Here, Newton almost certainly lied. The calculation was later found in his papers—and, as Newton may have recognized while Halley waited eagerly in his rooms, it contained an error.

No matter. Newton reworked his sums that fall, and then pressed on. In November, he sent Halley nine pages of dense mathematical reasoning, titled *De motu corporum in gyrum*—"On the Motion of Bodies in an Orbit." It proved that what would become known as Newton's law of gravitation—an inverse square relationship—requires that given certain circumstances, an object in orbit around another must trace out an ellipse, just as the planets of our own solar system were known to do. Newton went further, sketching the beginnings of a general science of motion, a set of laws that could, deployed properly, describe the how, the where, and the when of every bit of matter on the move anywhere—everywhere—in the cosmos.

The pamphlet was more than Halley had expected when he first goaded Newton into rethinking old thoughts. Once he read it, though, he understood immediately its larger significance: Newton hadn't just solved a single problem in planetary dynamics. Rather, Halley grasped, his friend had sketched something much greater, a newly rigorous science of motion of potentially universal scope.

Newton too grasped the opportunity before him. He was famously reticent, and he had published almost nothing for more than a decade. But this time he surrendered to Halley's encouragement, and began to write with the explicit intention of telling the world what he knew. For the next three years he developed a description of nature based on quantitative laws, applying those ideas to a whole range of problems of motion. As he completed each of the first two parts, he forwarded the manuscript to Halley, who took on the heroic double duty of preparing the dense mathematical texts for the printer while continually prodding Newton to get on with it, to deliver what he already knew would be the book of the age. Finally, in 1687, Halley received Newton's conclusion, the third section of the work, immodestly and accurately titled "On the System of the World."

This was the main event, nothing less than Newton's demonstration that his new science could encompass the universe. He took all the equations, the geometrical demonstrations, all the proofs he'd worked out to describe motion and produced a detailed, mathematically precise account of the behavior of the night sky, beginning with an analysis of the moons of Jupiter. He worked his way through the solar system, eventually returning home, to the surface of the earth. There he revealed a gloriously elegant result, an account of the way the gravitational tugs of the moon and sun produced the seemingly intractably complex action of the tides, turning the rise and fall of the sea into rigorous, calculable, scientific order.

He could have stopped there. It would have made sense, leading readers to rest at the natural end of one of the greatest stories ever told: an odyssey through the heavens above (those tiny, naked-eye-invisible motes circling Jupiter) to the earth below, our home, with every vista along the way accounted for by the workings of a handful of simply expressed laws.

There was, however, one more matter Newton chose to address before the last leaves of his manuscript could be released into Halley's hands. Comets had first brought Halley and Newton together: they had met after both had chased the bright comet of 1682—the one we now know as Halley's.* But in the last months of his work on *Principia,* a different object held Newton's attention: the Great Comet of 1680, discovered by the German astronomer and calendar maker Gottfried Kirch.

Kirch's comet was itself something of a milestone within the scientific revolution. On the night of November 14, 1680, Kirch had begun his regular night's work looking for something else entirely, mapping stars as part of a long-running observing program. That evening, he pursued his usual sequence: guiding his

* That object follows its own elliptical path—a much more elongated version of the orbits taken by the major planets—and it completes a single journey around the sun about every seventy-six years. Halley would later crack its true nature as a repeat visitor by analyzing historical sightings within Newton's gravitational mathematics in one of the early triumphs of the new science.

telescope to the first object of the night, taking notes, tracing the familiar patterns. Then his telescope shifted a little and something new appeared: "a sort of nebulous spot, of an uncommon appearance." He held on the stranger, tracking it long enough to be sure. It was no star. Rather, he'd found a vagabond, a comet—the first to be discovered using that icon of scientific discovery, the telescope.

For Newton the comet of 1680 offered a unique opportunity. He already knew the shapes of the planetary orbits he analyzed with his new mathematical laws—but this previously unknown visitor presented a novel challenge: could his universal gravitation account for motion no one had seen before? He set up his analysis by first plotting the path of Kirch's comet as revealed in reports from credible observers. He drew a line that connected each observation to reveal its track: a particular kind of curve called a parabola. Parabolas are mathematically kin to the ellipses traced out by the planets and moons Newton had just analyzed. The key difference: ellipses are closed curves; the earth, the planets, Halley's comet, NASCAR drivers* retrace their path with every trip round their oval courses. Not so any object on a parabolic path. Parabolas are open-ended, following lines that start out there, bend round a focus (the sun, for the comet of 1680), and shoot off again on a course that will never return to the old neighborhood.

Newton made sure every reader really, really understood that yes, the comet of 1680 rode a parabola in and out of the solar system. At the end of a very long and difficult book, he devoted

* For the record: the closed curved tracks on which NASCAR races are run generally are not perfect ellipses.

page after page to detailed lists of observations from all those comet hunters who had chased it through the constellations. He left nothing out—it was as if he wanted to cudgel his readers into silent agreement. By the end of his account no one could possibly doubt: the comet of 1680 roared in from who-knew-where, rounded the sun ... and then vanished into the unmapped vastness beyond, never, apparently, to return.

And then he performed one last feat. He extracted just three observations from his catalogue, three points on the comet's trajectory, and used his new mathematical model of force and motion to derive the orbit of that comet. He calculated, and the answer came back a perfect match: his results graphed onto that same course all those observers had found: a parabola.* Strip away the technical complexity—all those conics and curves and calculus masquerading as geometry—and what remained was the triumph, not just Newton's, but that of a whole new way of grasping the material world.

The account of the comet of 1680 gives his book its true climax. It was cosmic proof that the same laws that governed ordinary experience—the apple's fall, an arrow's flight, the moon's constant path—ruled all experience, to the limits of the universe. A parabola has no end nor beginning: One arm comes from the infinitude of the plane; the other arm shoots off to the same infinity. Placed in the material world, formed by the motion of a

---

* Newton later returned to the question of the true orbit of the comet of 1680, considering the possibility that it followed not a parabolic orbit, but instead a very elongated, long-period elliptical one. He was never able to derive such an orbit with confidence, though he believed it could have been a returning comet with a 575-year orbit. More recent analyses put the possible period on the order of ten thousand years.

comet swinging around the sun, the parabolic motion of the comet of 1680 traces out not just the events that take place in our immediate vicinity, but throughout the universe, from its deepest reaches and back out to them again.

Newton knew exactly what he had done. Near the end of the section on comets, he wrote: "The theory that corresponds exactly to so nonuniform a motion through the greatest part of the heavens, and that observes the same laws as the theory of the planets, and that agrees exactly with exact astronomical observations cannot fail to be true."

Edmond Halley agreed. Three years after he'd innocently asked for a single proof, he delivered to the printer the last pages of what Newton again immodestly, again accurately, titled *Philosophiae Naturalis Principia Mathematica*—The Mathematical Principles of Natural Philosophy. Getting Newton's enormous manuscript into book form while dealing with its ever-fractious author had left no time for Halley's own work since 1684, but now, at the finishing line, he granted himself his own victory lap. As *Principia* went to press, he exercised his editor's privilege to preface Newton's prose with a poetic assessment of the great work and its author: "But we are now admitted to the banquets of the gods/We may deal with the laws of heaven above; and we now have/The secret key to unlock the obscure earth; and we know the immovable order of the world/ ... Join me in singing the praises of Newton, who reveals all this,/Who opens the chest of hidden truth."

Hidden truths made plain! That was no poetic license. Amid all the talk of gods and heaven, Halley got it right. Newton had promised his readers the system of the world—and this is in fact

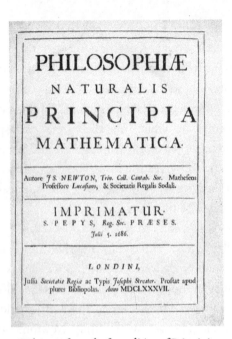

Title page from the first edition of Principia

what they received, a way to investigate matter in motion throughout the cosmos, to the utter limit of space and time. As the great French mathematician Joseph-Louis Lagrange famously said, "Newton was the greatest genius who ever lived, and the most fortunate; for we cannot find more than once a system of the world to establish."

Sir Isaac Newton died in 1727. Alexander Pope responded with his famous epigram: "Nature and nature's ways lay hid in night./ God said, 'Let Newton be,' and all was light." By the turn of the next century such apparent hyperbole would seem no more than predictable British understatement.

2

"A HAPPY THOUGHT"

March 1781, Bath.

It was William Herschel's day job that brought him to the city of Bath. He was a Hanoverian transplant, a musician by trade who had been made director of the Bath Orchestra in 1780. But if music paid the bills, the stars were his passion, ever since he saw the same view that has grabbed so many amateurs before and since: the rings of Saturn in all their glory.

Inspired, Herschel had taught himself how to construct a telescope (helped by his sister Caroline, reportedly the better of the two at the fine figuring of mirrors) and he had made the switch from stargazing to systematic astronomy as early as 1774. In Bath he settled into a seemingly prosaic program, an analysis of double stars. His goal: to distinguish these paired flecks of lights as either genuine examples of two stars in close proximity or as "opticals," two completely unrelated objects that just happened to fall together on a line of sight.

And so, on Tuesday, March 13, 1781, around the time women of the upper classes would have risen from their dinner table to allow their men to sit over their cigars and drink, he set himself to what had become a regular routine. He turned his largest and newest telescope—a Newtonian design with a 6.2-inch mirror, the best in England—toward a candidate double between the constellations Taurus and Gemini. One half of the apparent pair was utterly undistinguished, an ordinary point of light, just a

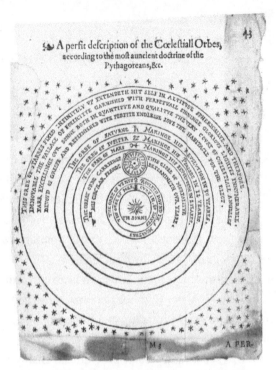

Thomas Digges's diagram of the Copernican cosmos, first published in 1576, depicted all the elements of the universe as it was known in the spring of 1781.

star. The other? It looked odd: fuzzy. Most important, it changed under magnification. Herschel recorded the anomaly in his entry for the night: "The lower of the two is a curious either nebulous star or perhaps a comet."

Herschel returned to observe the object repeatedly over the next month, until he was sure that it was indeed a candidate comet, a nearby object moving against a starry background. But by then it was clear that this "comet" was behaving oddly: it did not appear to be growing in the sky (or not much—Herschel persuaded himself for a while that he had measured an increase in its diameter) and it showed no signs of growing a tail. He reported his finding to the Royal Society, and other observers began to examine the object.

In May, two mathematicians—one French, one Russian—independently used the accumulating sightings to work out its true orbit. They proved (as Herschel had not) that this wayfarer was no comet. Rather, it traveled a nearly circular orbit, farther from the sun than that astronomer's gateway drug, the ringed giant Saturn.

From the beginnings of recorded history to that night in Bath, humankind had known exactly how many wandering stars—planets—traversed the skies. There were just six: Mercury, nearest the sun, then Venus, then our own Earth, Mars, Jupiter, and, most distant, Saturn. Even after 1609 when Galileo turned his strange new device, a tube with disks of figured glass at either end, and added Jupiter's moons to the solar system's family tree, that tally remained unchallenged. No longer. By convention, historians of astronomy date the discovery of the planet Uranus to Herschel's first glimpse on March 13, 1781.

Unsurprisingly, such an unprecedented find turned Herschel into a hero to his contemporaries. King George III offered him a £200 stipend if he would bring his observatory to Windsor Castle, then added a knighthood to sweeten the pot. Herschel's fellow astronomers took their reward too. Uranus created a unique opportunity, as it was the first major finding that could pose an independent test of Newton's mathematical version of reality. Put another way: a previously unknown object offered the astronomical community a chance to see how well their fundamental

*Pierre-Simon Laplace, in a posthumous
portrait by Sophie Feytaud*

tools actually accommodated not just what was known, but what had, until that March evening, remained unsuspected.

Among the first to take up the challenge was a young, brilliant French mathematician: Pierre-Simon Laplace. Laplace was something of a prodigy. Elected to the Académie Royale des Sciences eight years earlier, just twenty-four, he had since presented leading-edge work on pure mathematics, gravitation, probability, and more. When he heard the news of Herschel's observations, he immediately joined what was almost a stampede of European thinkers applying Newtonian ideas to the still unidentified object. Like Herschel himself, Laplace leapt to the conclusion that the object was a comet (which, of course, was hardly a crazy assumption: plenty of comets had been seen before and after the beginnings of telescopic astronomy, but no one had turned up any new planets—until then).

His attempt to calculate a plausible cometary trajectory failed, though, and he left it to others to expose Uranus's true nature. After that, though, Laplace took up the data again, and by early 1783 had come up with a novel, more general method for analyzing the motion of celestial objects. When he applied his new approach to Uranus, he produced the best description of its orbit yet achieved. To Laplace, that calculation was at once a rather minor display of analytical skill and one of the opening shots in what would become his life's work: using Newtonian physics, expressed in ever more sophisticated mathematics, to complete old Sir Isaac's foundational program, the construction in detail of a "system of the world" that could account for the behavior of every object in the universe, those known, and those yet to be discovered.

That work occupied Laplace for the next three decades and more. Between the 1780s and the first years of the new century, he created the most comprehensive account anyone had yet achieved of the interactions of the sun, its planets, their moons. With the rigor imposed by his increasingly sophisticated mathematical language, Laplace transformed the apparatus Newton used to show that the universe *could* be made intelligible into an epic narrative of how the universe actually behaves.

It wasn't always clear that the work would reach that happy ending. By the late eighteenth century, the dynamics of the solar system faced some open questions—several unanswered for decades. The most pressing: Jupiter was moving faster by the end of the seventeenth century than it had appeared to travel in earlier records, while Saturn seemed to have slowed down. The simplest analysis of the system (the sort Newton himself performed in *Principia*) implied that this couldn't happen. But manifestly it did, as documented by none other than Newton's dear friend Halley.

Enter Laplace, with a virtuoso demonstration of how his version of postrevolutionary science created new knowledge. Newton's gravitation boils down to this: an equation that tells you exactly how two bodies influence each other. If you know a handful of basic parameters—the masses of two bodies influencing each other, the distance between them—you can just turn the crank to figure out how much force each imposes on the other.*
Going from that force to a trajectory, an orbit, or the flight of a comet (or a cannonball) is a bit more complicated, but not much.

But such calculations are always idealized. Most of the time reality is messy, and the simplest applications of fundamental laws don't survive much contact with the world. The true test of Newton's science—of any abstract claim—comes when there is a conflict between existing understanding and some fact that doesn't fit. The failure to match the actual motions of Saturn and Jupiter with what the theory seemed to say should happen posed the question: what does that conflict mean? Is that a problem, or an opportunity?

---

* To find out how strongly the earth attracts the moon, for example, all you would have to do is plug in the numbers: the mass of the earth (almost 6×10^{24} kilograms, if you're wondering) times the mass of the moon (7.35×10^{22} kilograms, or roughly 80 times less than that of Earth) multiplied by the gravitational constant—6.67384×10^{-11} N m²/kg² (N is the symbol for newtons, units of force)—and divided by the square of the distance between the center of the earth and the center of the moon ($384,403$ kilometers—which is an approximation, as the moon's orbit is not perfectly circular). Do all the multiplication and division and you get your answer: the force of gravity exerted between the earth and the moon is about 1.99×10^{20}N.

Laplace held to his credo: "there is no truth in physics," he wrote, "more indisputable and better established by the agreement between observation and calculation than that all celestial bodies gravitate toward each other." This was, he added, Newton's doing, the outcome of the "most important discovery ever made in natural philosophy." The key to Laplace's adulation, though, lay with the necessity that observation and calculation *agree* on Newton's discovery. What to do, then, were they to disagree? As Laplace certainly knew, when the real world confounded theoretical explanation, that could simply mean that the theory was wrong. But there was another option. If something that *can* be measured doesn't fit, Laplace reasoned, the obvious next step is to look for something else, some other fact, perhaps some new way to understand the math itself, that could haul the real world back into agreement with its mathematical representation. Put another way: something out of whack suggests that there is something else out there to discover, maybe in nature, perhaps within the abstract ideas built to interpret nature's ways.

Laplace set to work on Jupiter and Saturn in 1785. He began on solid ground. The plain reading of Newton's laws said that Saturn and Jupiter should interact, and that the results of their gravitational dance could indeed be the *kind* of motion actually observed, the larger planet accelerating and the smaller slowing down. He reworked the calculation others had attempted before him and came to the same answer: the scale of the observed braking and speeding was about right, with a deviation small enough to support the intuition that the source of the error lay not with any failure by Newton, but in something Newton's heirs had missed.

With his own doubts thus relieved, Laplace next attempted

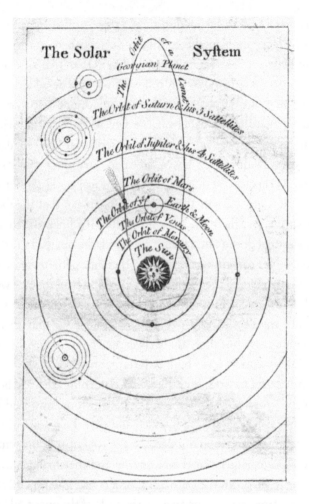

A map of the solar system published in 1791 as part of the Tom Telescope series of science books for children. In this very British setting, Uranus is still known as "the Georgian Planet"—an attempt at interplanetary nationalism that didn't last long.

what no one had yet been able to do: to construct a mathematical approach that would treat Jupiter and Saturn as a continuously changing series of related systems. Each change in the relative position of the two planets implied a different set of inputs for the equations that translate gravity into motion for each object. The "error"—that little bit of unaccountable extra acceleration of Jupiter—would, if this approach worked out, reveal itself as a perfectly "natural" outcome of the mathematics of gravity describing patterns of motion that evolve over time. This was a masterful bit of intellectual jujitsu, a shift from the observed behavior of celestial objects to the mental picture his fellow mathematical philosophers had constructed to model that behavior.

There was only one difficulty. Representing the two planets' positions and their relations to each other in three spatial dimensions and then allowing time to flow forced Laplace to construct a devilishly complicated system of equations. That mathematical model could only be solved (maybe) by an equally intimidating set of calculations. In the end, it took him a full three years—and that only with some extremely skilled help on the necessary analytical grunt work. But finally, in 1788, he was able to announce that he had cracked the mystery. The observed acceleration of Jupiter and deceleration of Saturn, he said, were caused by minute changes in the strength of the gravitational attraction between the two as their trajectories shifted. These changes occurred on a cycle that spanned centuries—929 years, in fact. The test of this claim turned on checking the exact timing of particular orbital conformations as far back as records could reveal—and in a test of theory against observation since 228 B.C.E., Laplace was able to show that the two planets obeyed Newtonian theory to the limits of the accuracy of available measurements.

It was a bravura performance, an almost ridiculously accomplished display of mathematical skill. It achieved something more too, a confirmation not just of Newton's theory as Laplace's "indisputable truth"—but of the truly revolutionary nature of the scientific revolution itself. Laplace had developed mathematical tools through which he expanded the reach of the fundamental laws Newton himself had created. This new math permitted a sharper description—more resolution in the picture, as it were—of the physical behavior it sought to model. Most important: that image wasn't simply more precise. It contained novel information, more detail; in this case, a hidden, slow dance of planets that takes nearly a millennium to unfold.

Thus the deep power of Newtonian science as Laplace and his peers understood it: it was an engine of discovery, powered by reason expressed in the particular rigor of mathematics. There was no end to the usual forms of exploration—the recent addition of Uranus to the solar system, for example, demonstrated that each advance in the technology of scientific instruments could reveal unsuspected terrain. But as Newton's followers and interpreters drove ever deeper into his mathematical reimagining of natural philosophy, it became clear that it was possible to explore *within* the math as well, journeys of the mind that could leap the page and guide explorers to discoveries in the wide world itself.

Another way of thinking about that transformation came with what Laplace did next. His monumental *Celestial Mechanics* fills five volumes—1,500 pages of dense analysis and calculation—and seeks to demonstrate that Newtonian universal gravitation, the mutual attraction of all objects in the sky on each other, could "when subjected to rigorous calculations, [yield] a complete explanation of all celestial phenomena and means of perfecting the tables and theories of the motion of heavenly objects."

By the end of his expansive calculation Laplace confirmed (to his own satisfaction) that he'd done it: the dynamics of the solar system—by implication, of the entire universe—were governed by the law of gravitation as Newton had first stated it. As his discovery of the roughly nine-hundred-year cycle of Saturn and Jupiter's motion had suggested, he now concluded that the solar system as a whole was stable, its motion conformable to itself on every time scale thus far examined. Such stability lent support to his third conclusion: the solar system—and by extension, the universe as a whole—was subject to what was formally called "determinism." Every event, everything that can be seen or measured or otherwise observed, was the outcome of some specific process or cause that could have generated only that outcome.

That claim contains an obvious implication—one that leapt out at Laplace's contemporaries. As the anecdote goes, Napoleon took a moment during the brief peace of 1802 to engage in a bit of intellectual banter. He entertained a few *savants*—Sir William Herschel himself, the distinguished physicist Count Rumford, his minister of the interior—a chemist by profession—Jean-Antoine Chaptal, and Laplace. After exchanging politenesses with Herschel, the First Consul next turned to Laplace, who had just published the third volume of *Celestial Mechanics*. Released from matters of state, Napoleon delighted in putting awkward questions to his guests, and so he told his mathematical friend that he had read Newton, and saw that his great book had mentioned God often. But "I have perused yours, but failed to find his name even once." Why is that, he asked?

In the grand tradition of this story, Laplace is reported to have replied, "I have no need of that hypothesis."

That sounds almost too perfect, but in an age of conversation as bloodless duel, it's not impossible that Laplace could have come up with such a perfect riposte on the spot. But even if the dialogue has been "improved," still, something along those lines passed between the two men. Herschel noted in his diary that Napoleon had asked "who is the author of all this" and that Laplace "wished to shew that a chain of natural causes would account for the construction and preservation of the system."

The deeper controversy concerns what Laplace really meant. Did he truly deny God's existence? Or was he saying something slightly more modest, that gods were irrelevant to the day-to-day management of reality? Such an underemployed deity could exist, and might even safely be regarded as an ultimate first cause, the source of the universe at the beginning of time. But after that, Laplace seems to have been saying, the divine need play no role in that universe's unfolding history. Newton had long since recognized that the mathematical principles of natural philosophy might tend a susceptible mind that way, but he denied the possibility. Rather, he saw in his studies of nature the chance to track down his God within creation; nature as it conformed to His will would reveal to an adept (like Newton) the hand of the deity in action. The uncertainties that he couldn't resolve in celestial mechanics only reinforced the idea that an all-powerful being still had a job to do to keep the entire system on track.

By the time Laplace finished solving the equations of motion for the solar system, though, his update to Newton's system appeared to run just fine on its own. In analyses spanning centuries, the planets needed no help to return to their courses. A "chain of natural causes" could account for Saturn's wavering orbit; for the motions of Jupiter's moons; for the demonstration of the long-term stability of all the planets' trajectories; even (speculatively) for the origin of the solar system as a whole. Laplace's God had indeed ceased to be a necessary actor; divine action becomes a "hypothesis"—and a superfluous one at that, not worth a moment's attention. As the historian Roger Hahn puts it, "Nowhere in his writings, either public or private, does Laplace deny God's existence. He merely ignores it."

That's a fair gloss on Laplace's view, but it's incomplete. Stripped to its essence, Laplace's career becomes a lifelong meditation on the question of cause and effect. Might it be possible to imagine that the tools of Newtonian science could yield perfect knowledge, a grasp of the full chain of events that led to any observable set of circumstances? Yes, he said, it is:

> We may regard the present state of the universe as the effect of its past and the cause of its future. An intellect which at a certain moment would know all forces that set nature in motion, and all positions of all items of which nature is composed, if this intellect were also vast enough to submit these data to analysis, it would embrace in a single formula the movements of the greatest bodies of the universe and those of the tiniest atom; for such an intellect nothing would be uncertain and the future just like the past would be present before its eyes.

That "intellect" is now sometimes called Laplace's demon. It is a mighty creature, to be sure, especially if you take its powers to the limit of Laplace's imagination. He wrote this description of his demon in 1814, the year Napoleon's empire shattered. Men, even or perhaps especially on a battlefield, are matter in motion. The intelligence that could trace the chain of cause and effect that carries each bullet to its halt, each soldier to his fate, could surely capture ("in a single formula") the collapse of the entire imperial cause.

And, as Laplace surely knew, *Celestial Mechanics* could be read as a kind of demonic text, offering a set of tools with which its readers could discover "the future just like the past" of the solar system. Such science doesn't merely describe, of course. The immediate successors to Newton's revolutionary generation used the interplay between meticulous observation and the mathematization of nature to generate *both* a formal account of what had already been measured, and what such measurements might suggest about what was yet to be observed. God-like knowledge was there—to be approached, if never fully attained.

Laplace died in 1827, seventy-eight years old. His analyses of the mechanics of heavenly bodies were already undergoing revision. Just as his own mathematical advances over Newton led him to a more comprehensive account of the solar system, so new methods enabled his successors to build increasingly precise models of planetary motion driven by universal gravitation. One man took pride of place in this ongoing transformation: Urbain-Jean-Joseph Le Verrier, who would fulfill his predecessor's vision of cosmic order with a discovery that seemed to his contemporaries the most perfect display imaginable of the power of Newtonian science.

3

"THAT STAR IS NOT ON THE MAP"

In the 1830s (and still) number 63 Quai d'Orsay turned an attractive face toward the river. In the guidebooks already being read by that novel nineteenth-century species, the tourist, number 63 is described as a "handsome house"—one, the writers warned, that concealed a much more plebeian reality. Visitors—by appointment only, no more than two at a time, welcome only on Thursdays—would be ushered into a courtyard, and then on to the rooms where workers, mostly women, took bales of raw tobacco through every stage needed to produce the finished stuff of habit: hand-rolled cigars, spun strands of chew that became "the solace of the Havre *marin*," gentlemen's snuff. Most of the campus was turned over to laborers serving the machines—choppers, oscillating funnels, snuff mills, rollers, sifters, cutters, and more. By the latter half of the nineteenth century, the works at the Quai d'Orsay would turn out more than 5,600 tons of finished tobacco per year, and was, according to the ubiquitous Baedeker, "worthy of a visit"—though indulging one's curiosity carried a price: "the pungent smell of the tobacco saturates the clothes and is not easily got rid of."

A spectacle, certainly, and as an early palace of industry clearly worthy of the guidebooks (themselves novelties). By any stretch of the imagination, though, the *Manufacture des Tabacs* was an odd place to look for someone who would become the most celebrated mathematical astronomer of his day—but not everyone follows a straight course to the person they might become. Thus it was that in 1833 a young man, freshly minted as a graduate of the celebrated *École polytechnique*, could be found every working day at the Quai d'Orsay, reporting for duty at the research arm of the factory, France's *École des Tabacs*.

No one ever doubted that Urbain-Jean-Joseph Le Verrier had potential: he had been a star student in secondary school, winner of second prize in a national mathematics competition, eighth in his class at the *polytechnique*. But his early career offered no hints to what would follow. Funneled into the tobacco engineering section in university, he was more or less shunted directly toward the Quai d'Orsay and the task of solving French big tobacco's problems.

It's not clear whether Le Verrier actively enjoyed the life of a tobacco engineer—or merely tolerated it. Nothing in his later career remotely suggests he was a born chemist. But he was consistent: if given a task, he got down to it. Never mind all that early training in abstract mathematics; if required, he could be as practical as the next man, and so turned himself into a student of the combustion of phosphorus. That was useful research—tobacco monopolists care about matches. But whether or not he relished his job, he certainly got out as soon as he could. A position back at the *École polytechnique* opened up in 1836 for a *répétiteur*— assistant—to the professor of chemistry. Le Verrier applied, and as an until-then almost uniformly successful prodigy, had every hope . . . until the post went to someone else.

Le Verrier would prove to be a man who catalogued slights, tallied enemies, and held his grudges close. But he never accepted a check as a measure of his true worth. A second assistantship became available, this time in astronomy. He applied for that too. Never mind his seven years among the tobacco plants; Le Verrier seems to have believed that he could simply ramp up his math chops to the standard required at the highest level of French quantitative science. As he wrote to his father, "I must not only accept but seek out opportunities to extend my knowledge. [...] I have already ascended many ranks, why should I not continue to rise further?" Thus it was that Le Verrier came into orbit around the great body of work left by that giant of French astronomy, Pierre-Simon Laplace.

Laplace had gone to his grave in 1827 convinced that he had solved the core of his great problem. To a pretty good approximation, he was correct. He had shown that the solar system as a whole could be rendered intelligible, its motions accounted for by Newtonian gravitation as expressed within mathematical models—"theories" of the planets. Properly employed, those models could describe the motions of the physical system explicitly, accurately, and indefinitely into the future. If there was some work left to do, new methods to be explored, more observations to be considered, discoveries within the system (like the newly discovered "minor planets"—asteroids—and comets), the basic picture seemed sound.

There were, though, more anomalies than the edifice of *Celestial Mechanics* acknowledged. Some of the theories of the planets were proving a bit less settled than Laplace had believed, and some, like Mercury's, were obviously inadequate, unable to predict the planet's behavior with remotely acceptable precision. Despite such problems (or possibilities), no researcher had yet returned to the whole of Laplace's program. Several astronomers in France and elsewhere worked on individual questions in planetary dynamics, but none were trying to resolve the system as a whole, to go from a theory of any given planet to one of the solar system, top to bottom.

Enter Le Verrier, of whom one of his colleagues would later say, "Laplace's inheritance was unclaimed; and he boldly took possession of it." Over his first two years at the *polytechnique,* Le Verrier surveyed the whole field of solar system dynamics, beginning to suspect that seemingly minor gravitational interactions might matter more than his predecessors had believed—that over time they produce effects that would be noticeable. He seized the opportunity, setting himself as his first major project the goal of recalculating at higher mathematical resolution the motions of the four inner planets—Mercury, Venus, Earth, and Mars. It took him just two years, a phenomenal pace given that he had started from zero as a mathematical astronomer.

Le Verrier presented his results to the French Academy of Sciences in 1839. He came to one striking conclusion: when you take one more term into consideration than prior calculations had attempted, it becomes impossible to say for certain whether or not the orbit of the inner planets would remain stable over the very long haul. Neither he nor anyone else knew how to find a complete solution to the equations that could confirm whether Mercury, Venus, Mars—and Earth—would remain forever on their present tracks.

Crucially, Le Verrier was already showing himself willing to tangle with the acknowledged master of celestial mechanics. Laplace had concluded from his studies of Jupiter and Saturn that the stability of the solar system was proved; here was a young man just two years into the field suggesting otherwise.

It was a fine first effort—good enough to garner attention from the men who could advance his career past an assistantship. At the same time, it was, as Le Verrier knew, still preliminary work, nothing more than recasting an old calculation. But it managed to hook him on celestial mechanics as a life project—and for his next major task, he set himself a problem that no prior researcher had been able to solve: Mercury.

If the planets were a family, Mercury would be the sneaky little sibling: it *might* be up to something, but it was so good at slipping past any attempt to pin it down it was hard to be sure. But that was no longer quite as true, as Le Verrier's gift for finding a ripe problem showed itself. Over the preceding decade, advances in instruments and technique made it possible to follow Mercury with a previously unattainable accuracy. He gave credit where it was due: "In recent times, from 1836 to 1842," Le Verrier reported to the Académie, "two hundred useable observations of Mercury have been carried out" at the Paris Observatory. With these and other records, he was able to construct a better picture of the way Venus influenced Mercury's orbit as the two planets moved from one configuration to another. That, in turn, led him to a new estimate of Mercury's mass, with his answer falling within a few percentage points of the modern value.

These were satisfying outcomes—filling in some of the more elusive details of one corner of the solar system. But Le Verrier really wanted a complete account of Mercury, a system of equations encompassing the full range of gravitational tugs that affect its orbit, which can be used to identify planetary positions past and future. Observations constrain such models: any solution to a model's equations has to at least reproduce what observers already know about a planet's orbit. More data meant more constraint, and hence a more accurate set of predictions about where the planet would go next. Those predictions, the "table" of the planet, are the test of any planetary theory.

The final exam for Le Verrier's first version of such a theory for Mercury came in 1845, its next scheduled transit of the sun, best viewed from the United States. Transits are ideal reality checks for such work: mid-nineteenth-century chronometers were accurate enough to note the instant Mercury's disk would cross the edge of the sun. On May 8, 1845, astronomers in Cincinnati, Ohio, watched as the clock ticked off to the moment Le Verrier had predicted for the start of the event. The astronomer at the eyepiece of the telescope trained on the sun saw "the dark break which the black body of the planet made on the bright disk of the sun." He called out "Now!" and checked his timepiece. Against Le Verrier's prediction, Mercury was sixteen seconds late.

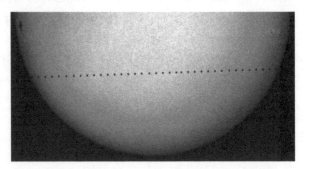

Mercury in transit across the face of the sun in 2006

This was an impressive result—better by far than any previous published table for Mercury, back to the one prepared by Edmond Halley himself. But it wasn't good enough. That sixteen-second error, small as it seemed, still meant that Le Verrier had missed something that kept the real Mercury out of sync with his abstract, theoretical planet. Le Verrier had planned to publish his calculation following the transit. Instead, he pulled the manuscript and let the problem lie for a time. Mercury would have to wait quite a while, as it turned out, for almost immediately he found himself conscripted into a confrontation with what was fast becoming the biggest embarrassment within the allegedly settled "System of the World."

Uranus was the troublemaker, and had been for decades. After Herschel's serendipitous discovery of the "new" planet, astronomers swiftly realized that others had seen it before, thinking it a star. In 1690, John Flamsteed, the first Astronomer Royal and Newton's sometime-collaborator, sometime-antagonist, placed it on one of his sky maps as the star 34 Tauri. Dozens of other missed-chance observations turned up in observers' records, until in 1821, one of Laplace's students at the Bureau, Alexis Bouvard, combined those historical sightings with the systematic searches that had followed Herschel's news to create a new table for Uranus, one supposed to confirm that it obeyed the same Newtonian laws that governed its planetary kin.

He failed. When he attempted to construct a theory of Uranus that could generate by calculation the positions observers had recorded since Herschel's night of discovery, he couldn't make the numbers work. Anything he tried that agreed with observations made since 1781 didn't line up with the rediscovered positions that had been misidentified as stars before that date. Even worse, when he focused only on the more recent, post-Herschel record, it quickly became clear that the planet was again wandering off course—or rather that reality and calculation diverged.

In the abstract, such uncooperative behavior might point to a very deep problem: if all the gravitational influences on Uranus had been accounted for, the failure to predict its motion would demand a reexamination of the theory behind such analysis. That is: it could threaten the foundations of Newton's laws themselves. One researcher, the German astronomer Friedrich Wilhelm Bessel, suggested exactly that, wondering if perhaps, just maybe, Newton's gravitational constant itself might vary with distance.

Such thoughts were thinkable, but horrifying. There was the worshipful awe the man himself inspired, of course, but more to the point, Newton's physics *worked*. The tides obeyed its rules; comets were brought to order under its provisions; cannonballs flew on courses perfectly described and explained by the exquisite logic of the *Principia*. Better, by far, would be any explanation that captured this seeming anomaly within a Newtonian framework.

It seems that, privately, Alexis Bouvard was the first to come up with a way to do so. In 1845, his nephew, Eugène Bouvard, reported to the *Académie* his own, unsuccessful attempt to bring Uranus's track to mathematical order. Following his uncle, he tried to resolve modern (post-Herschel) observations with older ones. He failed, and admitted as much. But still, he told his learned audience there was a way out, one his uncle had glimpsed two decades earlier. It was not the one Laplace had used to resolve the Jupiter-Saturn mystery. That involved improving the mathematical technique with which he attempted to describe the world out there. Rather, the older Bouvard reasoned, if all the known behaviors of the solar system could not account for the last residue of error—and crucially, if you maintained your faith in Newton—then the only remaining possibility was that something *unknown* would resolve the matter. Bouvard reminded his readers that if they imagined Uranus had remained undiscovered, then meticulous attention to Saturn would reveal the influence of some more distant unseen celestial body. In exactly the same manner, he wrote, it seemed "entirely plausible [to him] the idea suggested by my uncle that another planet was perturbing Uranus."

The Bouvards weren't the only ones to make that leap. By the early 1830s, several researchers began to think about the possibility of an object yet farther from the sun than Uranus. The older Bouvard had shared his notion with correspondents and visitors, one of whom carried the idea across the channel to England. One obvious difficulty kept this widening circle from doing very much with the idea, though. Uranus was too damn slow. Its eighty-eight-(Earth-)year-long period meant that systematic observations since Herschel had followed roughly half of a single journey around the sun. The Astronomer Royal George Biddell Airy conceded the plausibility of the idea of a trans-Uranian planet but quashed the hopes of one inquirer, writing that the mystery would resist solution "till the nature of the irregularity was well determined from several successive revolutions"—which is to say, only in that long run when all those then concerned would be dead.

Le Verrier disagreed. Or rather, his sometime-mentor, François Arago, the director of the Paris Observatory, thought that Herschel's planet had embarrassed astronomers long enough. Late in the summer of 1845, Arago pulled the younger man away from a brief dalliance with comets and, as Le Verrier recalled, told him that the growing errors within the theory of Uranus "imposed a duty on every astronomer to contribute, to the utmost of his powers." Le Verrier began by identifying several errors in the older Bouvard's sums. Those mistakes did not eliminate the unexplained wobbles in Uranus's orbit, so Le Verrier instead recalculated the planet's tables to define those anomalies as precisely as possible. With the intellectual ground thus cleared, he turned into a detective, seeking the as-yet-unidentified perpetrator that could have led Uranus astray.

As a good police procedural would have it, he soldiered on, examining—and eliminating—as many suspects as he could. Historian of astronomy Morton Grosser tallied Le Verrier's potential culprits: perhaps there was something about the space out by Uranus, some resisting stuff (an ether) that affected its motion. Was there a giant moon orbiting Uranus, tugging it off course? Might some stray object, a comet, perhaps, have collided with Uranus, literally knocking it from its appointed round? Le Verrier even paused on the fraught possibility that Newton's law of gravitation might need modification. Last: was there some as-yet-undiscovered object, another planet, whose gravitational influence could account for the discrepancies between Uranus's theoretically predicted and the observed track?

Le Verrier quickly rejected the first three potential candidates. He agreed with virtually every professional astronomer in thinking that modifying or rejecting Newtonian gravitation would be a final, desperate resort. Which meant that after several months of thinking about the problem, he was back to his prime suspect: an as-yet-undiscovered trans-Uranian planet.

With that, his task was sharply defined: once all the known

sources of gravitational influence were accounted for, what were the properties—mass, distance, finer details of its orbit—of the object that could account for the remaining anomalies in the motion of Uranus? In that form, the problem resolved down to a conventional problem in celestial mechanics, establishing and then solving a system of equations that described each of the components of the hypothetical planet's motion. Even so, given how little could be asserted with any confidence about the still hypothetical planet, then familiar or not, the task was deeply fraught.

Le Verrier first set up his calculation with thirteen unknowns—too many for someone with even his gifts to solve in any timely manner. So he simplified his assumptions. He argued that there had to be a sweet spot for at least some of the orbital parameters of the missing planet. As he would later write, it couldn't be too close to Uranus, for then its effects would have been too obvious. It couldn't be terribly far away, as that would imply a large enough mass to affect Saturn as well, and no such influence had been detected. He simply guessed that its orbit wouldn't be too sharply angled to the plane of the rest of the planets. He constructed a few more such arguments to fill in some of the gaps in the observational data from Uranus, which left him with a system of equations with just nine unknowns—which is to say, merely a hugely difficult operation, instead of an impossible one.

Calculating a unique solution within that model—one that would give him a prediction of the mass and position for the planet—proved almost ludicrously laborious. Being clever helped, as when he came up with a way to transform some of the essentially intractable nonlinear equations in the model into a larger set of linear expressions.* That made the calculation easier—possible, really—but at the cost of a horrific amount of grunt work to crank through the much greater number of steps the new approach required.

Even so, by the end of May 1846, Le Verrier had advanced to the point where he could report to the *Académie* that Uranus's orbit could be exactly described assuming "the action of a new planet"—and that it would be possible to show that "the problem is susceptible to only one solution ... there are not two regions in the sky in which one can choose to place the planet in a given epoch"—which was his rather grandiose way of saying he was nearing his answer. Near, but not yet all the way there: in this communiqué, he could do no better than suggest that the hypothesized trans-Uranian planet should lie in a region measuring about ten degrees across the sky.

Even that rather loose guidance was subject to a fair amount of uncertainty, too much to help anyone interested take a look. So Le Verrier returned to the mathematical grind, reworked his calculation, and on August 31, 1846, delivered an update: if anyone happened to have time to spare on a good telescope, they should find a planet beyond the orbit of Uranus at a distance of about 36 astronomical units,† visible about five degrees east of ∂ Capricorn—a fairly bright star within the Capricorn constella-

* A linear equation is one in which change in one variable produces a directly proportionate change in the result—producing a straight line when its results are graphed. In a nonlinear expression, a graph of the output produces a curve; systems of nonlinear equations are more difficult to solve than linear ones—often enormously so.
† The astronomical unit is roughly (and historically) the mean distance between the earth and the sun. (In modern astronomical units of measure, it is strictly defined as 149,597,870,700 meters, or about 93 million miles.)

tion. Its mass, Le Verrier declared, would be about thirty-six times that of Earth, and to the telescope-aided eye, it would reveal itself not as a point (like a star), but as a clearly discernible disk, 3.3 arcseconds in diameter.

And then, given that the cream of French astronomy had been told there was a new planet waiting to be discovered, what happened next?

Nothing.

The pursuit of Uranus's invisible companion is littered with in-hindsight incomprehensibly missed opportunities, but this one seems particularly hard to explain. Le Verrier had attacked Uranus at the direct request (almost an order) of the director of the Paris Observatory, at a time of nationalist competition on just about every axis imaginable, from the pursuit of empire to the pursuit of knowledge. He was acknowledged by his peers as the best analyst of celestial mechanics in France, offering up what would be (on the precedent of Herschel's fame after his accidental find) the discovery of a lifetime. And yet, none of his French colleagues could be bothered to point a telescope at the patch of night where, he had just told them, a career-making triumph could be theirs. It's true that the main telescope at the Paris Observatory was a mediocre instrument, and, as important, the astronomers there lacked the latest sky charts needed to check if anything that appeared in an eyepiece was already a known and distant star. Still, it remains odd that none of the astronomers who had first crack at Le Verrier's conclusions thought to try. All they risked was a night or two under the observatory dome, against a potential gain of a whole new world. But none did.

Finally, on September 18, Le Verrier lost patience with his compatriots. He wrote to a young German astronomer named

Johann Gottfried Galle. Galle had tried and failed to catch the more senior man's attention the year before, but now Le Verrier needed him. Some belated praise for Galle's research sweetened his plea: "Right now I would like to find a persistent observer who would be willing to devote some time to an examination of a part of the sky in which there may be a planet to discover." Galle received the letter five days later. Swallowing any lingering resentment for earlier neglect, he set to work that evening.

September 23, 1846, Berlin.

The night is quiet, very dark. Gaslights had come to Prussia's capital back in 1825, but there still weren't that many of them, and most were doused by midnight. After that Berlin belonged to those who cherished the night sky—among them, the watchers at the Royal Observatory, near the Halle Gate.

The "new" Royal Observatory in Berlin, depicted sometime after 1835.

This Saturday, Galle and a volunteer assistant, Heinrich Ludwig d'Arrest, command the main telescope. Galle stands at the eyepiece and guides the instrument, pointing toward Capricorn. As each star comes into view, he calls out its brightness and position. D'Arrest pores over a sky map, ticking off each candidate as it reveals itself as a familiar object. So it goes until, sometime between midnight and 1 A.M., Galle reels out the numbers for one more mote of light invisible to the naked eye: right ascension 21 h, 53 min, 25.84 seconds.

D'Arrest glances down at the chart, then yelps: "that star is not on the map!"

The younger man runs to fetch the observatory's director, who earlier that day had only reluctantly given his permission to attempt what he seems to have thought a fool's errand. Together, the trio continue to watch the new object until it sets at around 2:30 in the morning. True stars remain mere points in even the most powerful telescopes. This does not, showing instead an unmistakable disk, a full 3.2 arcseconds across—just as Le Verrier had told them to expect. That visible circle can mean just one thing: Galle has just become the first man to see what he knows to be a previously undiscovered planet, one that would come to be called Neptune, just about exactly where Urbain-Jean-Joseph Le Verrier told him to look.

Galle's sighting was the climax of what was almost immediately understood to be *the* popular triumph of Newtonian science. It's unsurprising, given the stakes, that the discovery of Neptune produced its share of controversy. The English astronomer John Couch Adams had followed the same reasoning as Le Verrier, performed commensurate feats of calculation, and had come to a very similar prediction at almost the same moment. However, he failed to persuade any of the astronomers at either Cambridge or the Royal Observatory at Greenwich to pursue a rigorous search. Still, a nationalistic priority battle followed, with British scientists pressing the case for Adams to receive co-discoverer credit with Le Verrier. That view held sway for more than a century, at least in the English-speaking world, though current historical analysis reserves pride of place for the Frenchman. A claim of discovery requires both the prediction and the actual measurement made on the basis of that prediction—and by that yardstick Le Verrier got there first.

Still, the fact that the priority battle was so important to the British astronomical establishment tells its own tale. The discovery of Neptune—driven by the mathematical interpretation of fundamental laws, so exactly as to reveal itself within hours of the start of the search—was recognized at once as both a stunning display of individual genius and a triumph for a whole way of knowing the world. In point of fact, Le Verrier (and Adams) had made several arbitrary choices to simplify parts of the problem, most notably guessing how far away the perturbing planet had to be. Those estimates were off by a lot, which would seem to undercut any claim of prescience for their theoretical calculations. And yet even such a miscue actually conveys something of the power of Le Verrier's reasoning, as he also figured out a way to frame the question to reduce the importance of distance for determining Neptune's position.* It wasn't luck (or not much) that delivered the new planet; it was the skill with which a very gifted Newtonian scientist had set up a calculation to tolerate a fair amount of wrongness in his assumptions. And in any event, for both the public and the world of professional astronomers, such slips simply disappeared in the glow of the bright, beautiful truth that Le Verrier had said go and look—there!—and you will see . . . and someone searched . . . and everyone saw.

* Le Verrier was able to take advantage of the fact that Uranus's track suggested that it and its unseen companion were in reasonably close conjunction in 1846, which meant a mistake in his estimate distance would affect the apparent position of Neptune much less than if the two planets were more widely separated.

That sequence transformed the discovery of Neptune from being merely spectacular (like Herschel's stumbling upon Uranus) into something more, a celebration of science as a whole. Le Verrier had confronted an uncomfortable fact, and then subjected it to theory, *the* theory, Newton's system of the world, to risk a prediction that then proved true. If ever there was a demonstration of how science is supposed to advance, here it was.

For Le Verrier himself Neptune was the golden ticket, his ride to the top of his corner of the world. Almost instantaneously it made him the most famous physical scientist in the world, and it pushed him up Paris's professional ladder at truly impressive speed. The romance of his victory, though, offered something more: validation for his faith, the belief that he (and humanity) could by pure force of intellect impose order on the natural world.

Interlude

"SO VERY OCCULT"

With Neptune in hand, one matter was settled. No working astronomer, no physicist, had any residual doubt about gravity. As Newton had described it, so it was: a *universal* force, at play throughout the cosmos, depending only on the masses contained within a system and on the inverse of the square of the distance between any two objects (in the simplest case). A century and a half of applying that law to ever more complicated arrangements had encountered no exceptions. Quite the reverse: with Neptune came the prospect of yet more to be discovered, as observers and the theoreticians honed their instruments and ideas.

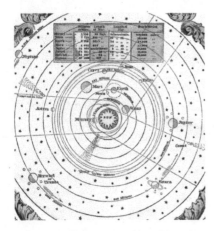

Such unbroken success vindicated Newton on another matter as well. He never publicly said he knew what gravity *was*. He refused to propose any specific notion to explain why one hunk of matter pulls on another. Such an account was, to him, unnecessary. He said so in one of his most famous

This nineteenth-century map of the solar system emphasizes the rigor with which Newton and his heirs ordered the cosmos.

one-liners, added in response to critics to the third edition of *Principia*: "I do not feign hypotheses"—or, in full, "I have not as yet been able to deduce from phenomena the reason for these properties of gravity and I do not feign hypotheses."

There you have it: one of the great non-apologies in the history of science.

It was a controversial gibe at the time—inflammatory, really, for all its august air of unconcern. Historians and philosophers still argue about what, exactly, Newton meant. But at a minimum, it's clear that Newton drew a sharp distinction between the way he thought natural philosophy should be done and how his opponents believed nature needed to be explained.

Here's the context. One of the most consistent requirements of pre-Newtonian natural philosophy comes in the demand for explicit determination of the causes, answers to the "why" and "how" for any phenomenon. From antiquity, this requirement led to explanations like the one Aristotle gave of the mechanism of planetary motion: the planets ride on rotating spheres, he said, which themselves move through eternity thanks to a shove from the original source, the prime mover. In medieval reworkings of that idea, God takes over from Aristotle's formless author, but the concept of a direct connection between motion and a mover remains. See, for example, a gorgeous image found in the fourteenth-century manuscript *Breviari d'Amor* (an Abstract of Love): two angels, elegant in their robes of green, seated outside the sphere of the fixed stars, turning deep blue cranks. Those divine agents become, in the words of the artist Michael Benson, an image of "changeless supernatural beings winding the clockwork of temporality."

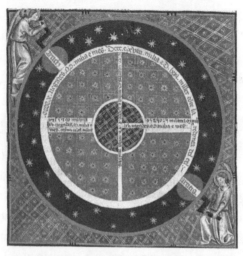

In the cosmic order depicted in Matfre Ermengau's fourteenth-century manuscript Breviari d'Amor, *everything beyond the moon is pure and perfect, and the machinery of heaven drives the ceaseless round of the sublunary sphere.*

By Newton's time such divine engineers had surrendered to more purely inanimate drivetrains, but the need to provide a direct account of cause and effect remained. Thus, when René Descartes set out to create a modern cosmology, he suggested that space had to be full of some mysterious fluid in which the motion of celestial objects could be driven by whirlpools, vortices that could impart the necessary impetus to the planets. That notion solved what was seen as the essential problem. Here was machinery that could be imagined to make a cosmos go.

Unfortunately for such a mechanistic explanation, Newton showed in *Principia* that vortex physics was wrong both in particular (the mathematics Descartes developed to describe his vortices failed to predict planetary positions correctly) and, much more important, was unnecessary. Once you allow a handful of axioms—gravity acts according to an inverse square law; it does so everywhere; and the force imparted by gravity to an object moves that object in accordance to three simple laws of motion—then that's it.* You don't need anything else to provide an accurate account of falling fruit here on Earth or the tracks of moons, comets, and planets through the night sky. This gravity is disembodied, abstract, something Newton called a force, without ever quite defining the term, and without saying just how that force imparts its impulse to whatever it touched. No levers here, no gears, no mover, prime or not. Instead, action at a distance, something that seemingly leapt from one mass to another, incorporeal, instant in its effects.

This is what so offended Newton's critics, themselves no slouches as natural philosophers and mathematicians. To them, Newton had abandoned the direct "local" explanations of Cartesian physics (and Aristotelian, for that matter)—the way that such explanations brought cause directly into contact with effect right where any effects occur. Once he denied the demand to explain *how* nature worked, he undermined (seemingly) the very nature of physical explanation. Gottfried Leibniz, the nearest Newton had to an intellectual equal, complained publicly that absent an explanation for how it made things go, Newton's theory verged on blasphemy: "without any mechanism ... [gravity] is an unreasonable and occult quality, and so very occult that it is impossible that it should ever be done though an angel or God

---

* These laws are simple in the sense the physicist Richard Feynman meant when he described solving a problem in Newton's *Principia:* " 'Elementary' does not mean easy to understand. 'Elementary' means that very little is required to know ahead of time in order to understand it, except to have an infinite amount of intelligence."

himself should undertake to explain it." To Leibniz as to many of his contemporaries, the argument that gravity was somehow just the way nature did it was an unacceptable surrender. Newton's view was, in their eyes, a bizarre unwillingness to accept what seemed obvious. If a planet orbits the sun in a path that suggests a tug between the two bodies, then it must be the natural philosopher's job to find out what "really" connects the two.

But what if that necessity was an illusion? Newton's refusal to assert what he did not know was more subtle than a simple rejection of mechanical dogma. Instead, the deeper truth hidden within Newton's seeming intellectual modesty comes from the realization that there is a real gap between mathematics and physics. Newtonian theory exists in mathematical form, in equations. There, gravity is simply a quantity, some number of units of force that can be calculated in a context of other quantities: the mass on which the force acts; the acceleration experienced by that mass. There is no need to invoke a specific set of physical connections. The test of such a relationship, of a claim about the motion of bodies, is to observe, to measure, and to match calculation to what can be seen.

That's how Newton's math became physics: an abstract relationship like force is equal to mass times acceleration:

$$F = ma$$

Or, for the gravitational force between two bodies:

$$F = \frac{G * m_1 m_2}{d^2}$$

—enables anyone to figure out where Mars would be next Tuesday.

"I feign no hypotheses" may mean many things, but at the least, it says this: the mathematical form of physical law is its own hypothesis, a proposition about the material world, subject to the judgment of measurement and observation. Once it meets that test, such an *un*feigned hypothesis gets woven into the fabric of reality; it becomes, as Newton had proclaimed, an explanation that "cannot fail but to be true."

Looking back, finding Neptune so precisely where calculation said it should lie may seem merely one more in the line of confirmations of the most successful scientific idea in history. But that's only part of the story. The objections Newton faced in his lifetime did not simply disappear on his say-so, that regal "I Do Not Feign!" The heart of the Newtonian revolution lay with the claim that a purely mathematical argument was a sufficient account of events in the physical world—all of it, the full, unmeasured sweep of the heavens and our own mundane experience here on Earth: the same laws governing a ball dropping from a child's hand or the tide sweeping away a sand castle to Neptune appearing first on the page and then in the eyepiece of Galle's telescope. But the conviction that equations do in fact represent reality did not triumph in a day.

It didn't even for Newton himself. He wasn't—he couldn't be—truly a Newtonian, utterly convinced of the sufficiency of mathematical explanations. Like anyone, he was a man of his time and place—and there was as much (or more) in him of the past that made him as of the future he helped make. He was a secret alchemist, committed to esoteric studies to uncover how change happens in nature. That alchemical spirit infused his ideas about gravity, including how the material fact of a planet could, disembodied, reach across open space to move another body. He saw God's hand at work throughout the solar system, the entire cosmos—no abstract spirit but the ultimate agent in the material world. Newton's heirs downplayed, and then simply ignored, this side of their hero's convictions. (The University of Cambridge would even decline to accept a donation of Newton's alchemical papers as "of very little interest in themselves.") Instead, thinkers like Euler and Lagrange and Laplace and finally Le Verrier constructed in Newton's name a worldview in which mathematics and not mechanism became *the* scaffolding of the universe—math, without God—to the point where Laplace's "I have no need of that hypothesis," could both echo and overwhelm Newton's "I do not feign..."

These Newtonians extended Newton's ideas to more and more complicated problems. They vindicated them again and again throughout the solar system until, at last, Neptune crushed all doubt. Le Verrier's calculation signaled not just advance but victory: the approach to nature established by Newton and developed since was no mere set of clever tools. Rather, it offered the definitive account of how the cosmos actually works.

It is for this reason that Newton is remembered not simply as a great thinker for his day, but as the greatest scientist ever. For all the secrets he kept, the private thoughts he harbored—for all the (to our eyes, not his) crazy, almost magical beliefs that informed his natural philosophy—the legacy of "I feign no hypotheses" remains. It is the cornerstone of the scientific account of material experience: rigorous observation and measurement of the physical world, expressed and analyzed in the language of number.

Part Two

NEPTUNE TO VULCAN

(1846–1878)

4

THIRTY-EIGHT SECONDS

Success mellows some. Not everyone, though, and certainly not Urbain-Jean-Joseph Le Verrier. The man who discovered Neptune "at the tip of his pen," received—and very rapidly came to expect—a hero's reception. His professional peers understood what he had done, while the public received him with all the reverence due a magician who could conjure a planet out of equations. As the mathematician Ellis Loomis wrote in 1850, "the sagacity of Le Verrier was felt to be almost superhuman. Language could hardly be found strong enough to express the general admiration." Loomis himself was slightly less moved, noting that the outcry was "somewhat extravagant." No matter, he added. Even a more sober assessment of Le Verrier's achievement would still have earned him, Loomis concluded, "the title of FIRST ASTRONOMER OF THE AGE."

Le Verrier concurred. He was first among not-quite-equals, as he emphasized in what was almost his first professional act after Galle's sighting was confirmed by other observers. At issue: what to call the new planet. Le Verrier had an obvious answer, given the naming convention for the rest of the planets, drawn from the Roman pantheon (with Uranus the one Greek outlier). He proposed Neptune, god of the sea and Jupiter's brother. That choice tangled the sequence of the family tree—Saturn was both Jupiter and Neptune's dad and Uranus was Saturn's father. Still, Le Verrier's choice fit with the broad sense of how to address a respectable planet, the same sentiment that attached names like Ceres and Pallas to the largest of asteroids discovered earlier in the century.

So far so good, though there was some dissent from the English, who preferred Oceanus in a nod toward the claim that an astronomer from their sea-girt island had a hand in the discovery. But while that move may have irritated Le Verrier—with cause—it seems to have occurred to him rather quickly that he might have undersold his triumph. So he withdrew from the naming stakes and turned to his colleague François Arago, director of the Paris Observatory, to represent him at the new planet's christening. Arago did so, making a proposal that might have aroused some suspicion in uncharitable minds: Le Verrier's planet should be called . . . Le Verrier!

The honoree made an unconvincing show of humility, a public stance undercut by his sudden shift on what to call the seventh planet from the sun. Now, for the first time in his career, he took to addressing Uranus by the name only English astronomers still (occasionally) used: Herschel. As for Britain in the eighteenth century, so for France—and M. Le Verrier—in the glorious nineteenth. The maneuver failed (obviously), in part because Herschel's son John rejected the idea of relabeling his father's find, and more because no astronomers beyond Paris, and not many there, could stomach a celestial Le Verrier glowering down on them night after night. The earthbound version ultimately gave up the attempt, and the consensus remained with what had from the start seemed like the obvious choice: Neptune.

Still, plenty of rewards did come his way. As Herschel had before him, Le Verrier won the royal notice, receiving the *Légion d'Honneur* from the hands of Louis Philippe of France. More practically, from this moment in his career he began to gain real power in and ultimately dominance over the French astronomical establishment. Just months after the discovery, French officials asked him to submit a proposal for his future research. In reply, he proposed to out-Laplace the master himself, to "encompass in a single work the entire ensemble of the planetary system ..." Such a project would, he wrote, "reconcile and render everything harmonious, if possible, and when this cannot be done, to declare with certainty that there exist causes of perturbations still unknown, whose origins are then and only then revealed."

Le Verrier himself had no illusions of the scale of the project. Consider, he told the ministry, what was involved: first, he would have to gather a comprehensive catalogue of observations of each planet in turn. Then came the building of the system of equations that could account for every known influence on the individual objects under scrutiny, one by one, on the Neptune-validated faith that Newton's universal gravitation governed all such interactions. Next, with the observational data fully incorporated into a mathematical

Urbain-Jean-Joseph Le Verrier, as the French public at mid-century knew him.

model of the planet, he (and his assistant) would calculate the planet's table—the specific numbers that predict its position at any arbitrarily chosen time. Finally, with all the planets thus rendered on paper, it would at last be possible to see whether any measured motion in the observed solar system escaped Le Verrier's written account. If any such an anomaly appeared, there the next Neptune would lie, trembling on the point of discovery.

In all, Le Verrier estimated, it would take at least a decade to complete the task, probably more—and meeting even that deadline would require the services of an assistant to take on the monotonous labor of calculation along with his own complete freedom to pursue his inquiry for as long as it would take.

The excellencies at the Ministry of Public Instruction raised no objections—and why would they? One cheaply paid assistant and the license to do research for which he was clearly suited was hardly a difficult request to grant the man who conquered Neptune. Perhaps predictably, though, Le Verrier didn't start right away. He had applause to reap—a tour of England in 1847 was only one such distraction—and Paris between 1848 and 1850 was roiled by the political transition that ultimately produced France's Second Empire, with the original Napoleon's nephew seizing power as Napoleon III. Le Verrier, like Laplace before him, took part in revolutionary politics—and like his intellectual ancestor, managed to navigate treacherous shifts in power unscathed. By 1850, with his position stable and the power struggle resolved, he could once again focus on problems in celestial mechanics. Arrogant ass though he may have seemed—and been—to a growing number of his peers, he quickly demonstrated that he was, in fact, the first astronomer of his age.

If Le Verrier had a secret superpower—some special gift that propelled him to insights his contemporaries missed—it lay with his knack for sniffing out the subtleties of the physics implied by his calculations. As good as he was as a mathematical astronomer, none of his peers would have seen him as the best mathematician of his day. Nor was he an observational maestro. When he became director of the Paris Observatory in 1854, he managed men who stood to the eyepieces, but it wasn't his job (or inclination) to do so himself. Instead, his signal contribution came in reasoning from equations to their solutions to an interpretation of what the numbers implied about the actual events in the world. He was able to remind his peers of this as soon as he stepped off the carousel of fame and returned to the concentrated labor of celestial analysis. He tackled a seemingly minor problem—a close examination of the orbits of what were then still known as the minor planets, the asteroids. At stake was a question of origins: how could astronomers explain the growing heap of rubble found in the gap between Mars and Jupiter?

The first (and, unsurprisingly, the largest) asteroid, Ceres, had been discovered in 1801. The next year, Pallas was identified by the German astronomer Heinrich Olbers. Olbers published the first guess as to why there was more than one object sweeping through orbital tracks that, in the rest of the known solar system, would belong to just one much larger object, a "major" planet. His idea: the two little objects identified so far were what was left from a planetary catastrophe: once, he argued, there must have been a much larger body out there—a hypothetical fifth planet from the sun.

That planet would be dubbed Phaeton—Apollo's son, struck

down by Zeus (Jupiter) after he lost control of his father's sun-chariot. Olbers suggested that the asteroids were all that were left after the celestial Phaeton had been destroyed in an early solar system cataclysm. Later notions included the idea that Phaeton had passed too close to Jupiter, to shatter under the stress of the gravitational force imposed by the largest planet in the solar system, or that it had been whacked by another large object at some earlier moment. Whatever the particular history involved, Olbers made one clear prediction: if that missing planet had once existed, only to be blown apart, then there should be dozens, hundreds—who knew how many?—asteroids to be discovered in the same general area that had already yielded up Ceres and Pallas.

He was quite right about that, of course, as he proved by becoming the first observer to discover a second asteroid, Vesta, in 1807. It's also important to note that his idea that the early solar system experienced some kind of a demolition derby can't be dismissed out of hand. After all, the best current reconstruction of how our Earth got its moon is the so-called giant impact hypothesis: an object about the size of Mars colliding with the proto-Earth one hundred million years or less after the birth of the solar system as a whole. At least one such hypothetical whack-and-hover object even has its own name, Theia (for the Titan whose daughter was Selene, goddess of the moon).*

---

* There are a number of variations on the currently most widely accepted idea of the collision hypothesis for the formation of the moon, and there is at least one proposal that suggests no collision took place at all. But the core idea explains key issues raised by the discovery of similarities between the composition of the earth and of the Apollo moon rocks and by the dynamics of the earth-moon system. So current betting is weighted toward some version of a cosmic wreck between the early Earth and some other large body.

At the same time, there's a common trick nature plays on its would-be investigators: resemblance, the human urge to map the unknown onto the already known, can be a snare. Just because something *looks like* something else doesn't mean that the backstory for both must be the same. Rocks scattered across the sky may appear to be a rubble field left behind by an explosion . . . but unless you stop to think how else you might get there, you rely on assumptions not in evidence. Olbers couldn't escape his sense of the familiar. Le Verrier would.

Le Verrier's first brush with the asteroids had come a decade earlier, when he determined that Jupiter and Olbers's own Pallas had orbital periods locked in an 18/7 ratio—a gravitational resonance analogous to the one Laplace had earlier found linking Jupiter and Saturn. Now, returning to the minor planets, he rejected Olbers's hypothesis. There was no need, he argued, to invoke a catastrophe. Instead, taking the formation of the asteroids as merely another example of the process that gave birth to the rest of the planets, he made two predictions: first, even though the catalogue of *"petites planètes"* listed just twenty-six asteroids with known orbits, his view coincided with Olbers: there should be a "prodigious number" more, begging to be discovered as soon as observers got their hands on better instruments. As those new bodies swam into view, he reasoned, it would become possible to determine their true distribution across the night sky. In doing so, he argued, observers would find evidence for his claim that "the same cause that united the material in each of the principal planets had also arranged [*distribué*] the smaller bodies into distinct groups."

Le Verrier was right. The distribution of asteroids found since he proposed his idea reflects the primordial process of planet formation—the accumulation of particles of material into first smaller objects, then rocks, then "planetismals." Out to the orbit of Mars, the sequence continued up to the accumulation of the major rocky planets. In the asteroid belt, though, Jupiter's gravitation whipped the accreting objects up with enough violence to prevent any single large object from forming. Instead, as Le Verrier predicted, the asteroids do form into groups and families that are connected by common orbital dynamics—a clumping driven by Jupiter's gravitational influence—though Le Verrier himself did not correctly identify Jupiter's role in that result. But for any later correction, Le Verrier here displayed the key capacity for scientific advance: he saw past the easy similitude—who doesn't love a wreck!—and chose not to reason backward from appearances.

Instead, he made the critical assumption: a new phenomenon does not necessarily demand its own new cause to account for its sudden appearance. Observations are essential; but, as Le Verrier argued through his analysis of the minor planets, they are not in themselves sufficient. The scientist's duty confronting some new circumstance is to find the meaning within the flood of new data. Half a century later, his compatriot the great mathematician Henri Poincaré would put it like this: "We can not know all facts and it is necessary to choose those which are worthy of being known."

That sounds reckless, arrogant—who are you or I to conclude some detail is "worthy"? But no, says Poincaré. There is an internal logic, a way of framing the beauty in nature that removes the scientist's particular caprice from the process. The trick was to lay claim only to those facts that could "complete an unfinished harmony, or ... make one foresee a great number of other facts." Le Verrier confronting the asteroids found in the facts already established—twenty-six well-mapped orbits—a path to such elegant efficiency. Look for more asteroids, he said. Place them within the deeply established apparatus of Newtonian gravitation, and use that analysis to enlarge the already well-ordered taxonomy of the planets. For Poincaré, scientific thinking at its best was an artist's performance; Le Verrier, confronting the asteroids, delivered work to please the most exacting connoisseur.

More success didn't mellow the man. Le Verrier turned out to be a viciously effective academic politician. By the early 1850s, he laid his sights on control of the Paris Observatory, and with it control over the most significant astronomical research program in France. Against Le Verrier stood François Arago, the incumbent director and his onetime patron. The two had fallen out in the late 1840s, when Le Verrier made his first attempt to maneuver behind the scenes to gain control of at least part of the Observatory's resources. Arago held on, but as he grew ill in 1853, he and his allies set up a committee to review possible successors. Le Verrier hadn't wasted his time either, and had accumulated enough influence with the French government to protect himself. The Ministry of Public Instruction ordered a halt to the search process, and instead set up a review of all facets of the observatory's operations by a commission that included Le Verrier. The committee's final report, with Le Verrier's hand obvious in its recommendations, described an institution fallen behind the times, burdened with obsolete equipment, poorly located, and ineffectively led. Clearly a new approach and a new man was needed— and the report proposed the creation of a permanent directorship whose authority, it was stated, "must be absolute ... not to be inhibited or compromised by the intervention of a deliberative body." The Ministry agreed, and in an order delivered ten days after receiving the committee's work, named the obvious candidate as the new and all-powerful leader of France's astronomical ambitions: Urbain-Jean-Joseph Le Verrier, of course.

The Paris Observatory, as a newspaper illustrator glossed it in 1862

Those who knew him best braced for trouble. Joseph Bertrand would serve for more than two decades as the permanent secretary to the Academy of Sciences. He observed Le Verrier at close range for years, and he wrote that almost as soon as Le Verrier became Neptune's discoverer, he played poorly with others. He remembered incidents dating back to those very first years of celebrity in which, "he showed little curiosity regarding the work of anyone else; he corrected others on occasion, and highlighted their errors, never softening the harshness of his manner in such encounters.... Through each dispute, the admiration he received at first did not last."

Rising from colleague to boss didn't improve Le Verrier's behavior. He fired all those he felt had been too close to the last administration—so remorselessly that one biographer suggests he drove one man to suicide. He was no kinder to the subordinates he hired himself. Camille Flammarion, an assistant who joined the Observatory staff in 1858, recalled him as "haughty, disdainful, inflexible ... [an] autocrat [who] considered all the employees at the Observatory as his slaves." Charles Aimé Joseph Daverdoing, the artist who had painted his portrait in the first glorious afterglow of Neptune, knew the private Le Verrier as "a good-natured fellow, very cheerful and good company." But he confirmed that at work, "Le Verrier was excessively demanding ... [and] did not make allowance for the age or the stamina of the workers. ... He was never one to bite his tongue, and once or twice laid hands on someone." This was more than mere gossip. The personnel records from his tenure as director reveal a monstrous casualty list: during Le Verrier's first thirteen years in the job, seventeen astronomers and forty-six assistants abandoned the Observatory.

But even if Le Verrier failed the test of power, his own abilities were never in question. Once he had completed his initial coup at the Observatory, he set out to make good on his earlier promise to complete a full theory of the solar system. He still had some preliminary work to do—most significantly analyzing several thousand measurements of the sun's (relative) motion across the sky. In 1852, when he began tackling the problem, the prevailing best estimate of the distance between the earth and the sun was 95 million miles (about 153 million kilometers). By 1858, Le Verrier was able to correct that figure by more than 2.5 percent—a huge improvement—to yield a measurement of 92.5 million miles, impressively close to the modern number of 92.995 million miles (or 149.597 million kilometers).

There's an air of routine in such data crunching—the kind of dry record keeping needed to fix the third or fourth decimal places of navigational tables to keep OCD ship captains happy. Certainly the luckless and overstretched assistants who performed the in-the-trenches work likely felt they were strapped to an endless assembly line, tabulating position after position, and then powering through the endless sums to yield the precision Le Verrier required. But in fact, this analysis was absolutely crucial to the larger goal of resolving all the remaining anomalies in the tables for each of the eight planets. With his much improved account of the sun, Le Verrier next sought to recast the tables for the four inner planets, to bring them to the same level of accuracy already achieved for the giant outer planets that had (among much else) yielded up Neptune in the first place. When he did so, he found that one simple change—altering upward estimates of the mass of Earth and Mars—combined with the new solar distance, allowed him to make sense of three of the four bodies to be explained. The mathematical representations of Venus, Earth, and Mars all behaved properly. The calculated version of each planet produced a chart that matched the one formed by the observational record of the three material planets making their way around the sun.

One, though, obstinately refused to conform: Mercury.

Mercury, of course, was an old adversary. Recall that in the 1840s, it had eluded Le Verrier in his first attempt to construct a mathematical model of its behavior. His had been the most accurate to date, but he understood the implications of its near miss on the timing of Mercury's 1845 solar transit. Better-than was not good enough; it still wasn't right.

Back then, Le Verrier acknowledged that there was no obvious solution to the problem. He wrote, "If the tables do not strictly agree with the group of observations, we will certainly not be tempted into charging the law of universal gravitation with inadequacy." Why not? Neptune, of course, with its once-and-for-all demonstration of the power of Newton's theory, or, as he put it, "these days, this principle has acquired such a degree of certainty that we would not allow it to be altered."

Instead, Le Verrier argued, such an error must be due either to "some inaccuracy in the working [calculation] or some material cause whose existence has escaped us." At the time he wasn't sure where the fault lay. Given the complexity of analysis and the relative poverty of data on Mercury, "we will not be able to decide," he wrote, whether to blame "analytical errors or . . . the imperfection of our knowledge of celestial physics."

There the matter rested. It wasn't until 1859, sixteen years after his first attempt, that he found himself free to return to the problem. He was forty-eight years old, at the height of both his fame and, by all witness testimony, his mathematical powers. He had the resources of the Paris Observatory at his disposal. Mercury's theory should have been a straightforward task.

It was . . . and it wasn't. The older Le Verrier had one absolute advantage over his younger self: better data. He reexamined the information he had used in 1843—measurements of Mercury's motion made at the Observatory itself. To that he added the best observations it was possible to make at the current state of astronomical technology: transits, with high-quality records for Mercury extending back to 1697. With a good clock and an accurate fix on where on earth the event was being viewed, timing a planet's entry or exit from a transit ranked among the most precise measurements available to astronomers.

Le Verrier launched his assault following his usual plan. First he mapped out Mercury's actual orbit with all of its components of motion as described by the empirical data: direct measurements of Mercury's behavior. Next came calculation: what do Newton's laws predict for Mercury, given all the known gravitational contributions of the planets as well as the sun? Any discrepancies—astronomers call them "residuals"—between the empirical picture and the theoretical one must then be explained. If there were none, then the theory of the planet was complete, and the model of the solar system would be one step closer to being done.

But there was a leftover result. It was a small number—tiny, really—but the gap between theory and the data was greater than estimates of observational errors could explain, which meant the problem was real. That settled one matter: it strongly suggested that Mercury's difficulties almost certainly lay not with flaws in Le Verrier's analysis, but rather in something unknown out there in space.

The particular anomaly he found is called the precession of the perihelion of Mercury's orbit. In the squashed circle of an elliptical orbit, the point at which a planet comes closest to its star is called its perihelion. In an idealized two-body system, that orbit is stable and the perihelion remains fixed, always coming at the same point in the annual cycle. Once you add more planets, though, that constancy evaporates. In such a system, if you were to map each year's track onto a single sheet of paper, you would over time draw a kind of flower petal, with each oval just slightly shifted. The perihelion (and its opposite number, the aphelion, or most distant point in the orbit) would move around the sun. When that shift comes in the direction that the planet moves in its annual journey, the perihelion is said to advance. As every schoolchild confronting geometry knows, a circular (or elliptical) orbit covers 360 degrees. Each degree can be divided up into sixty minutes of arc; each minute into sixty arcseconds. Le Verrier's analysis told him that this was happening to Mercury: its perihelion advances at a rate of 565 arcseconds every hundred years.

Next came a round of celestial bookkeeping: how much of that total could be explained by the influence of the other planets on Mercury. Venus, as Mercury's neighbor, proved to be doing most of the work. Le Verrier's sums revealed that it accounted for almost exactly half of the precession, 280.6 seconds of arc per

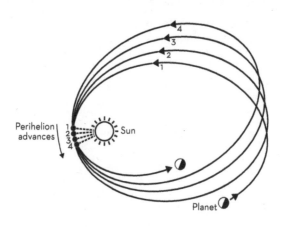

This is an exaggerated view of Mercury's orbit. The perihelion shift, repeated over decades, produces a flower petal design around the sun.

century. Jupiter provided another 152.6 to the total, Earth 83.6, with the rest causing scraps of motion. The total: 526.7 arcseconds per century.

A century and a half later, the one irreducibly extraordinary fact of this work remains how incredibly small an "error" Le Verrier uncovered. The unexplained residue of Mercury's orbital dance came down to a perihelion that landed just .38 seconds of arc ahead of where it should every year. To put it into the form in which Le Verrier's number became famous: every hundred years, during which Mercury travels a radial journey of 36,000 degrees, the perihelion of its orbit shifts about 1/10,000th beyond its appointed destination, an error of just 38 arcseconds per century.

Tiny, yes. But the excess perihelion advance of Mercury retained one crucial property: it wasn't zero. Le Verrier knew what such unreconciled motion must mean. If Mercury moved where no known mass existed to push it, then there was some "imperfection of our knowledge" waiting to be repaired.

5

A DISTURBING MASS

Le Verrier was hardly infallible, to be sure, but there were some errors he simply did not commit. Mercury's orbit does precess around the sun. It does so at a rate that cannot be fully accounted for by any combination of gravitational influences within the solar system. Le Verrier's number for the residual motion of Mercury—38 arcseconds per century—is a little off the modern value of 43 arcseconds, but he got it as nearly right as anyone could in 1859, given the limitations of the data at his disposal. Le Verrier never doubted the work. Nor did his fellow astronomers. For them, it was in fact fantastic news: the unexplained invites discoveries.

Of all men, Le Verrier knew what came next: in his book-length report on Mercury, he said as much: "a planet, or if one prefers a group of smaller planets circling in the vicinity of Mercury's orbit, would be capable of producing the anomalous perturbation felt by the latter planet. . . . According to this hypothesis, the mass sought should exist inside the orbit of Mercury.' "

Le Verrier then took the next step, figuring out how big an intra-Mercurian planet would have to be to drive the perihelion advance. Assuming it lay roughly halfway between Mercury and the sun, he wrote, its mass would have to be about the same as its neighbor. That posed a problem, as he well knew. If it were that big, why hadn't anyone seen it yet? Even if a Mercury-sized planet in the predicted orbit would usually be hidden within the glare of the sun, "It must be unlikely," he wrote, that it could avoid detection "during a total eclipse of the sun." Thus, Le Verrier proposed an alternative: "a group of asteroids [*corpuscles*] orbiting between the sun and Mercury."

That conclusion must have seemed a bit deflating to Le Verrier's readers. Adding to the lengthening list of minor planets, even in such an exotic location, hardly stacked up to finding Neptune. But the stakes of the search were just as high in both cases. Until Mercury's precession could be accounted for, the anomaly represented a violation of the cosmic order, unthinkable (of course) to all of Newton's heirs. Hence Le Verrier's urgency: "It's likely that some of these [asteroids] will be sufficiently large to be seen on their transits across the disk of the sun. Astronomers, already engaged with all the phenomena that appear on the surface of that star will without doubt find here another reason to track any spot they may see, no matter how small." In other words: all those sunspots you folks have been tracking? Some of them might be little planets. Go get 'em!

For those unwilling to wade through the long job of sorting sunspots, there was one other way to speed discovery. Le Verrier had published a short form of his Mercury findings in the September 12, 1859 edition of the Académie's proceedings, *Comptes Rendus*. In the same issue, the secretary to the Académie, Hervé Faye, wrote that the best chance of seeing Le Verrier's hypothetical asteroids was during a solar eclipse. By good fortune, the next readily accessible eclipse was almost upon them, to come on July 16, 1860, visible over northern Africa and Spain. During totality, the region closest to the limbs of the sun would suddenly be freed from the brutal glare of the sun, until "at the decisive moment," Faye wrote, the few minutes of totality "would suffice to explore much of the area designated by M. Le Verrier."

Faye's report sparked a wave of preparation. Locations were chosen—near Bilbao, perhaps, or a few miles west of Zaragoza, or maybe across the Mediterranean to a point on the coast around Algiers—wherever each observing team believed they would find the best chance of clear skies on the sixteenth of July. Given Le Verrier's history, it seemed plausible that one or more little planets might appear, even on a first attempt. And maybe they'd already been seen! Recalling the tally of misidentified sightings of Uranus, Le Verrier's announcement sent some back to old records, looking for anything that might qualify as an intra-Mercurian body since Galileo had first turned his telescope skyward back in 1609. No persuasive candidates materialized in this first pass—but then again, despite Herschel's turn of luck, knowing what you're looking for is a powerful aid to discovery. On to Spain!

Edmond Modeste Lescarbault was a humble, almost diffident man. He lived a small life, confined mostly to a modest compass between the Seine and the Loire rivers, about seventy miles west and a touch south of Paris. He had studied medicine, and in 1848 opened a practice in a little country town, Orgères-en-Beauce. He stayed put there for the next quarter of a century. He died in 1894, ninety years old, locally honored—the street where he kept his surgery is now named rue du Dr. Lescarbault—and generally forgotten.

The country doctor had one great passion. As a boy, he had fallen in love with the night sky. Children grow up, of course, and most put away childish things. Not Lescarbault. Like many before and since, he discovered in astronomy the same consolation that would later comfort Albert Einstein: the contemplation of "this huge world, which exists independently of us," which, he wrote, serves as "a liberation."

For Lescarbault, liberating himself from the daily medical round led him to build a genuinely impressive amateur's observatory: a low stone barn with a modest dome at one end. There he mounted a perfectly competent telescope, a four-foot-long refractor with an objective lens almost four inches in diameter. He would steal time there between patients, just minutes sometimes, sneaking from his office to the dome to look, perhaps to dream, just a little. The discovery of the asteroids in the belt between Mars and Jupiter led him to wonder: where else might such treasures lurk? An answer came to him on the 8th of May 1845—the day Le Verrier missed the timing of Mercury's encounter with the sun.

Lescarbault watched Mercury's moving dot across the solar face, untroubled by any mathematical subtleties. Instead, he thought not about the planet in transit, but whether there might be other unobserved transits to seek. If a Ceres- or a Pallas-sized asteroid lurked close to our star, its transits would likely be the only opportunity to see it—and the search for such events would be a perfect target for an enthusiastic amateur astronomer, eager for the thrill of finding something in the cosmos that not one other human in all of time had perceived.

He was slow to act on that epiphany. Ordinary life intervened. His medical practice needed nurturing, for one thing, but more important, he was a true amateur. He lacked both the knowledge and tools to achieve the precision needed to capture a phenomenon as delicate as an asteroid breaching the limb of the sun. It took him more than a decade to prepare, but by 1858, he had fitted his telescope with homemade instruments good enough to fix the position of objects within its field of view. He was, at last, ready to go hunting.

Saturday, March 26, 1859. Orgères, on the edge of spring, enjoys a sun-warmed afternoon. The flux of patients eases. As is his habit, Dr. Lescarbault takes the opportunity to retreat to his observatory. He turns his telescope toward the sun. An object leaps into view: a small, regular dot, just inside the edge, or limb of our star. He makes an estimate of its size: about one quarter the apparent diameter of Mercury. He has just missed its first appearance at the edge of the sun. Working backward from its apparent rate of motion, he estimates the time it crossed the solar limb at almost exactly four o'clock or, to be precise, at 3h 59m 46s P.M., plus or minus five seconds. He writes that down, using a piece of charcoal to scratch on a board. Another patient arrives and, likely with unrecorded frustration, he pulls his eye from his telescope. A few minutes later, he returns. The spot is still there, moving across the face of the sun. He tracks it continuously now, noting its nearest approach to the center of the solar circle, and then the instant and place it disappears over the solar limb. He records the time again: 5h 16m 55s. Total transit duration: one hour, seventeen minutes and nine seconds. If an asteroid were ever to be discovered within the innermost wards of the solar system, this is how it would reveal itself. Lescarbault meticulously transcribes his notes, and then . . .

Does nothing . . .

For nine months . . .

Until, at last, he permits himself to write a letter to be delivered—by hand—to Paris.

He "broke his silence," Le Verrier later wrote, "solely because he had seen an article in the journal *Cosmos* on [my] work on Mercury." Lescarbault described the data he had collected that Saturday in March—and added one bold claim: "I am persuaded also that [the planet's] distance from the Sun is less than that of Mercury, and that this body is the planet, or one of the planets, whose existence in the vicinity of the Sun M. Le Verrier had made known a few months ago, by that wonderful power of calculation which enabled him to recognize the conditions of the existence of Neptune...."

Lescarbault entrusted it to a M. Vallée, "Honorary Inspector General of Roads and Bridges," for delivery to the obvious recipient, Le Verrier himself. Dated December 22, 1859, it reached Paris a few days later. Le Verrier's first reaction—as he told it—was one of doubt. But he was prepared to hope. There was only one way to be sure if Lescarbault could possibly have made the observations he claimed to have achieved: meet the man; inspect his instruments; *test* him. No matter how unlikely it might be that some rural hobbyist could have plucked such a prize, even the possibility that he might made any delay intolerable. Le Verrier was promised to his father-in-law's for a New Year's Day celebration—but the train schedules showed that it was just possible that he could get to Orgères and back to Paris before midnight on the 31st. He commandeered Vallée to return with him as a witness, and the two men set out to see if Lescarbault's "planet" might actually exist.

Le Verrier and Vallée arrived at Orgères-en-Beauce unannounced, covering the last twelve miles from the nearest railway

Dr. Lescarbault's observatory-turned-tourist attraction in this postcard from 1863.

station on foot. A few days later, he painted for the Académie a calm, almost placid picture of the encounter: "We found M. Lescarbault to be a man long devoted to the study of science.... He permitted us to examine his instruments closely, and he gave us the most detailed explanations of his work, and in particular of all the circumstances of the passage of a planet across the sun." The two men from Paris made Lescarbault walk them through each phase of his observation until they were convinced that their amateur had in fact seen what he said he had—and, crucially, that his interpretation of the event was correct. "M. Lescarbault's explanations, the simplicity with which he offered them to us gave us total conviction that the detailed observation he had completed must be admitted to science."

Le Verrier told the story very differently in private. Released from the conventions of scientific discourse, he seems to have composed a hero's epic. Abbé Moingo, editor of the same journal, *Cos-*

mos, in which Lescarbault had first read of the problem of the precession of Mercury, was present at one of these performances. Le Verrier told of setting out for Orgères, Moingo wrote, assuming that no mere rural medico could have both discovered a new planet and kept quiet about it for nine months. Yet he had "a secret conviction that the story might be true." At the doctor's house, the astronomer confronted "the lamb" who trembled before the lion from Paris: "One should have seen M. Lescarbault... so small, so simple, so modest and so timid." Le Verrier roars; Lescarbault stammers—and yet, according to the Abbé, still manages to defend himself at every turn. "You will then have determined ... the time of first and last contact?" Le Verrier demanded, noting that measuring first contact is "of such extreme delicacy that professional astronomers often fail in observing it." Lescarbault admitted that he had missed first contact, but had estimated the timing by checking how long it took for his spot to travel the same distance again it had already passed from the limb. Not good enough, said Le Verrier, and on learning that the doctor's chronometer lacked a second hand, stormed "What! With that old watch, showing only minutes, dare you talk of estimating seconds? My suspicions are already too well founded."

Lescarbault rallied from even that devastating assault, though, showing his visitors the pendulum he used to count seconds, and reminding the astronomer that as a doctor "my profession is to feel pulses and count their pulsations ... I have no difficulty in counting several successive seconds." By this point in the remembered (and, to modern ears at least, suspiciously dramatic) account, it's becoming clear what Moingo (and/or Le Verrier) is doing. The ebb and flow of leonine attack, each swipe seemingly fatal, and yet disarmed by a counter from the charmingly naive lamb, enlarges Lescarbault. The famous astronomer plays the part of the skeptic (never mind how much he may have hungered for one outcome over another), while the country doctor becomes more and more a competent, even an excellent man of science. The interrogation lasted an hour, enough to exhaust Le Verrier's reservoir of doubt. At the last, he surrendered: "with a grace and dignity full of kindness, he congratulated Lescarbault on the important discovery he had made." He would lead Lescarbault to a more tangible reward as well, securing within the month the *Légion d'Honneur* for "the village astronomer" who had, it seemed, discovered the first intra-Mercurian planet.

The next step was all Le Verrier. Lescarbault had none of the mathematical skill needed to transform his observation into a planetary orbit. Le Verrier did so in less than a week. By making the assumption that its orbit was nearly circular, he calculated that the new planet would complete one revolution around the sun in just under twenty days, on a path that never exceeded eight degrees distance from the sun. Such an object would be difficult if not impossible to see directly. But if Le Verrier's analysis were even close to correct, the proposed planet would repeat its transits two to four times each year.

With that, planet fever hit the popular press—*The Times* of London, *Popular Astronomy* in the United States, *The Spectator* (which had some very kind words for Dr. Lescarbault). Alternative orbits were proposed: one reexamined the data on the assumption that the new planet traced a highly eccentric ellipse around the sun. Others returned to old records to see if Lescarbault's planet had been seen and ignored previously—and just as with Uranus and Neptune, candidate objects soon turned up, reaching double figures in a series of sightings stretching back to the mid-eighteenth century.

It was clear more work needed to be done, beginning with a repeat observation of the mystery object. Nonetheless, the celebrations continued heedless of any lingering uncertainty—and for good reason. The faith in the new planet stood in equal measure on Le Verrier's own reputation and the rock-solid logic behind the discovery. Mercury's perihelion precession was and is real. Newtonian gravitation provides an obvious solution to such a problem. The appearance of an object exactly where necessity suggested it ought to be made perfect sense. It fit. It had a moral right to be true.

Celestial facts need labels. This time, there was no nationalistic controversy to navigate, no tussle pitting "Oceanus" vs. "Le Verrier." The common practice held: planets major and minor took their identities from the gods of antiquity. It's an oddity of history that there is no record of who first fixed on the ultimate choice, but the decision was easy. A body that never escaped the intense fires of the sun had only one real counterpart on Olympus: Venus's husband, the lord of the forge. By no later than February 1860, the solar system's newest planet knew its name:

Vulcan.

6

"THE SEARCH WILL END SATISFACTORILY"

Vulcan's career began happily. Weeks after Le Verrier's announcement, no less an old rival than the Royal Astronomical Society bowed before the new planet: "The singular merit of M. Lescarbault's observations will be recognized by all who examine the attendant circumstances; and astronomers of all countries will unite in applauding this second triumphant conclusion to the theoretical inquiries of M. Le Verrier." More practically, the news evoked the sincerest form of flattery—claims of prior, never-recorded encounters with the newcomer. Benjamin Scott, Chamberlain of the City of London and an avid amateur astronomer, wrote to *The Times* to assert that he had long before found an intra-Mercurian planet: a candidate object the apparent size of Venus glimpsed at sunset "at or about Midsummer 1847."

Scott's "discovery," reported only in a conversation with a fellow of the Royal Astronomical Society, could hardly be taken seriously, but working astronomers wondered if they too had missed the prize. Rupert Wolf, a Zurich-based astronomer long fascinated by sunspots, reviewed his own and other solar observations to find potential mistakes—Vulcan transits he may have mistaken as mere spots—and came up with twenty-one possibilities that he published, and sent directly to Le Verrier as well, highlighting four that seemed the closest match to Lescarbault's object.

Wolf's list caught the attention of another astronomer, J.C.R. Radau, who used the data from two of Wolf's candidates to refine what could be extracted from just a single Vulcan sighting. Radau joined other professionals who sniped at "the procrastinated publication of Dr. Lescarbault's remarkable observation." But once past his pique, Radau performed his analysis meticulously, generating exactly what astronomers needed to attempt the next phase of Vulcan research: a prediction for an observable transit. With the assumption that Wolf's two suspects were in fact that same object as the one Lescarbault had seen, Radau published the results in early March: transits of Vulcan could next be expected between March 29 and April 7.

Radau's transit would be visible in the southern hemisphere, and astronomers there readied themselves for the moment of discovery. The director of the Victoria Observatory, a Mr. Ellery, monitored the sun at half-hour intervals. Major Tennant, head of the Madras station, went one better, reporting that "the sun's disk was watched every few minutes from March 27 to April 10." At the Sydney Observatory, Mr. Scott set up a parallel search. Ellery summed up the outcome for all three: the planet hunt performed by multiple observers reached the end of the predicted period for Vulcan transits "without success."

That was a blow, but hardly a fatal one. It had been obvious from the start that Vulcan would be hard to observe. If it weren't, any large body—Mercury-sized or thereabouts—would have been seen long since. That was why Le Verrier had thought that an intra-Mercurian asteroid belt was the more likely option until Lescarbault's report had raised hopes for a singular Vulcan. Still, while Lescarbault's object appeared to be bigger than most if not all asteroids, his notes suggested it would be as small as one twentieth the diameter of Mercury. At that scale it could not account for all of the perihelion advance Le Verrier had discovered. Lescarbault himself largely disappeared from view after his sudden burst of fame. The *Légion d'Honneur* he received in 1860 did not change his habits; he remained a country doctor and amateur astronomer until his death. After Le Verrier's visit, he made no further claims about any intra-Mercurian objects.

Working astronomers, though, still had to deal with the problem. Any calculation of Vulcan's orbit based on one or a few sightings would be an approximation at best, and stood a good chance of being just wrong. For Le Verrier as for many of his peers, the missing transit expected on the basis of Radau's calculation only demonstrated, once again, that doing astronomy at the limit of the math and empirical capacity is really, really hard. The necessity of the search hadn't changed one bit: Mercury still precessed, and whatever was compelling it to do so remained to be found.

As it was, swiftly.

In the middle of the nineteenth century, Manchester, England, prided itself on being smart as well as rich. In 1861, the city showcased both its wealth and brains as it hosted Britain's largest celebration of knowledge, the annual meeting of the British Association for the Advancement of Science. Charles Darwin had published *The Origin of Species* less than two years before, and that explosion continued to reverberate through every gathering of the learned. At the Manchester meeting Darwin's defenders prepared to battle religious doubters. One speaker, the "blind economist" Henry Fawcett, made the ultimate claim: Darwin was a true scientific hero, one who solved his problem by the same methods, the same approach to experiment, observation, and generalization that the great Isaac Newton himself had used in his physics.

Much else was discussed, of course—advances in dredging engineering, a report on birds of New Zealand, news from the balloon committee. The astronomy section was relatively quiet, but all in all, the meeting reflected a basic truth about Victorian curiosity: it was ubiquitous, constant, the common passion of both professionals and amateurs. No wonder, then, that Manchester's citizen-scientists would chase new planets.

So it happened, on the morning of March 20, 1862, a "Mr Lummis, of Manchester" stole a few minutes to peer at the sun through a small telescope. As the formal report in *The Astronomical Register* told it, Lummis was watching "between the hours of 8 and 9 A.M., when he was struck by the appearance of a spot possessed of a rapid proper motion." The object was startling enough that Lummis called for a witness, and they "both remarked on its sharp *circular* form." Lummis tracked the spot for twenty minutes before being called in to his day's work. By the time he returned to his telescope, the object was gone, "but he has not the slightest doubt of the matter." Radau and a colleague repeated the by-now familiar exercise, constructing the elements of an orbit from incomplete observations, and they found that Lummis's potential Vulcan was at least compatible with Lescarbault's, even if there wasn't enough data to settle the matter once and for all.

There were doubters. Two professional astronomers, the American Christian H. F. Peters and the German Gustav Spörer dismissed Lummis's "discovery" as a mere sunspot. But for many others, Le Verrier among them, the ongoing identification of plausible Vulcans, in sightings that allowed for at least rough estimates of consistent trajectories, made an ultimate validation seem inevitable. By the mid-1860s, *The Astronomical Register* itself seemed to view the matter as settled, listing Vulcan (without stating whether it was Lescarbault's object or some other) as the innermost body in its "Descriptive Account of the Planets."

Matters soon grew more complicated, though. Reports of sightings continued to arrive, some from reputable observers, others from unknowns. In 1865, an otherwise completely obscure M. Coumbary wrote to Le Verrier with a detailed account of an observation he made in the city that he—an unreconstructed Byzantine, apparently—referred to as Constantinople. With his telescope in Istanbul he watched as a black spot separated itself from a group of sunspots and appeared to move independently. He continued to track the object for forty-eight minutes, until it vanished over the limb of the sun. Le Verrier endorsed Coumbary's report, noting that though he didn't know his correspondent, his information seemed to him to be marked by a combination of "exactitude and sincerity." In 1869, a group of four eclipse mavens at St. Paul's Junction, Iowa (one a lady, as contemporary records took pains to mention), saw "*with the naked eye* what they termed a little brilliant at a distance about equal to the Moon's diameter from the Sun's limb"—an object that at least two others (one equipped with a small telescope) seem to have noted as well.

To those for whom the logical necessity of Vulcan was overwhelming, this spray of messages was comforting, not proof in and of itself, but an ongoing accumulation of information building on an already established pattern. The lack of a pure Neptune moment must have been frustrating, but given the inherent difficulty of the problem, such momentary glimpses gained significance each time another letter from some sincere and precise stranger reached Paris. As *The New York Times* put it, "a little scrap of positive evidence overbears an immense amount of negative." But despite a growing heap of such hopeful wisps, Vulcan remained almost maliciously elusive when confronted by a systematic search.

Benjamin Apthorp Gould had a perfect Boston pedigree: son of the headmaster of the Boston Latin School, grandson of a Revolutionary War veteran, he graduated from Harvard College—where else?—in 1844, all of nineteen years old. Then, having paid his debt to ancestry, he kicked over the traces. Heading to Europe, he took work at the Greenwich, Paris, and Berlin observatories just as Neptune made its (perceived) solar system debut. He studied math at the University of Göttingen, and in 1848 became the first American to receive a Ph.D. in astronomy—still only twenty-three! On returning to Boston in 1849, he was appalled by the primitive state of research in his home country, and took it on himself to transform American astronomy. Most important for the future of the discipline as a whole, in the 1860s he became one of the first investigators skilled in the new technique of astrophotography, the marriage of a camera to a telescope.

Gould brought his cameras with him when he traveled to observe the same 1869 eclipse at which the amateurs had spied a possible Vulcan. He set up in the town of Burlington, Iowa, working on the right bank of the Mississippi River. His goal: to study the solar corona—the sun's atmosphere, visible only during totality—and to survey the region close to the sun as precisely as possible, looking for whatever might reveal itself within the orbit of Mercury. He and his assistants made forty-two photographs

CLEAR FOR ACTION.

*Astrophotographers from the Canadian Eclipse Party, at
their observing station in Iowa in August 1869.*

during the eclipse. Gould also examined many of what he estimated were four hundred images made by others along the path
of totality. In all those pictures, he saw—nothing.

Gould sent his findings to Yvon Villarceau at the Paris Académie. He began with a baseline estimate: in the shadow of the
eclipse, a planet or planets substantial enough to account for
Mercury's motion should shine about as brightly as Polaris, the
North Star, a second magnitude object—easily seen by the naked
eye.* His photographic equipment, Gould wrote, was sensitive

---

* The magnitude scale for celestial objects dates back to Greek astronomers, and was
based on a rough visual distinction between the brightness of different stars. It was
originally calibrated on Polaris, which was assigned a magnitude of exactly two.
Smaller (and negative) numbers are brighter. The sun as seen from Earth has a magnitude of −26. In the modern definition, a magnitude one star—like Antares or
Spica—is one hundred times brighter than the magnitude of six objects that lie at the
edge of human naked-eye perception.

enough to detect any object down to the limit of unaided human perception, well below what he considered the plausible threshold for the discovery of Vulcan. Thus, he concluded, "I am convinced that this investigation dispenses with the hypothesis that the movement of the perihelion of Mercury results from the effects of one or many small interior planets." I've looked, he said, and Vulcan ain't there.

Not so fast, though: Villarceau added a note of his own to the published version of Gould's letter. It wasn't necessary to accept the American's conclusion as absolute, he argued. There were configurations of asteroids, for example, that could both provide the necessary gravitational influence on Mercury *and* evade detection. In other words: the problem remained. Mercury still wobbled, and in Newton's cosmos, its motion still demanded something like a Vulcan. Absence of evidence, to invoke what has become a cliché, could not be taken as evidence of absence.

Others agreed. William F. Denning was by general agreement Victorian Britain's greatest amateur astronomer. He had made his reputation with the first comprehensive analysis of the motion of the Perseid meteor shower, still to be seen from late July to its peak in mid August, and meteors remained his primary obsession. Vulcan, though, was a sufficiently pressing problem to draw his attention. He was an obligate organizer, and he used his influence to launch a systematic search for solar transits during the next likely window: March and April of 1869. He persuaded fifteen other sky-watchers to put the sun "continually under observation, when visible . . . with a view of rediscovering the suspected intra-Mercurial planet Vulcan."

Vulcan obstinately refused to appear.

Denning tried again the next year, recruiting a team of twenty-five to chase the elusive planet during the spring transit season in 1870, and yet once more with a plea to collaborators in 1871. As he gathered his volunteers, he had declared that his aim was to settle the issue once and for all. "There is every reason," he wrote, "to suppose that the search will end satisfactorily, if not successfully." End it did. After three conscientious attempts at locating the missing planet, he seems to have concluded that there was nothing more to be done. He did not repeat his call for aid on the search, and those fellow amateurs of the sky who had responded to him were released to their prior ambitions.

After what was to that point the largest systematic search for the object since word of Lescarbault's sighting first spread, Denning's null result left Vulcan in a predicament. An explanation for Mercury's errant motion remained necessary. On one side of the ledger, there was the blunt fact of Le Verrier and his genuine abilities. No one doubted his calculation, and no one should have—a restudy of Mercury's perihelion advance in the 1880s confirmed and slightly enlarged the very real anomaly he identified. Glimpse after glimpse of possible candidate planets offered tantalizing hints—yet a decade into the search, the most rigorous observers kept coming up empty. What could be done?

A way out was obvious to the more mathematically sophisticated Vulcan hunters. People simply could have gotten their sums wrong. There were enough imprecise assumptions about the elements of a putative Vulcan's orbit so that calculations for transits could just be wrong. Princeton's Stephen Alexander told his fellow members of the National Academy of Sciences that he had reworked Vulcan's elements to arrive at the conclusion that there should be "a planet or group of planets at a distance of about twenty-one million miles from the sun, and with a period of 34 days and 16 hours." In other words: we may have been looking in the wrong places, or at the wrong times. Vulcan could be elusive, but not absent.

That claim seemed to be confirmed when Heinrich Weber— for once, an actual well-trained professional astronomer—sent word from northeast China that he had seen a dark circular shape transit the sun on April 4, 1876. Sunspot expert and Vulcan devotee Rupert Wolf passed word of his colleague's sighting on to Paris, taking a bit of a victory lap as he did so. He told Le Verrier that "the interval between Lescarbault's observation and Weber's amounts to exactly one hundred and forty eight times the period" that Wolf had calculated so many years before.

The news enthralled Le Verrier—and energized yet another corps of planet seekers more eager than expert. As historian Robert Fontenrose put it, "everyone with a telescope was looking for Vulcan; some found it." For a time, *Scientific American* eagerly trumpeted each new "discovery": from "B. B." in New Jersey to a Samuel Wilde in Maryland, to W. G. Wright in San Bernardino, to witnesses from beyond the grave, in the form of a minister who remembered that Professor Joseph S. Hubbard "had repeatedly assured him he had seen Vulcan with the Yale College Telescope." New Vulcans kept turning up that autumn in seemingly every mail delivery, until at last *Scientific American* cried "Uncle!" and, following its December 16, 1876, issue, declined to publish any more such happy memories. It was as if the question of Vulcan had ridden a seesaw since 1859. Occasional sightings and seemingly consistent calculations would propel it up to the top of the ride; hard-nosed attempts to verify its existence sent it crashing back down. Now, for all that the editors of *Scientific*

American had tired of the flood of anecdotes, the teeter-totter was pointing up: between the one seemingly authoritative report from China and the sheer number, if not the quality of sky-gazer accounts, the matter of Vulcan seemed just about settled.

The popular press certainly thought so. In late 1876, *The Manufacturer and Builder* said, "Our text books on astronomy will have to be revised again, as there is no longer any doubt about the existence of a planet between Mercury and the sun." That autumn, *The New York Times* was even less bashful, interrupting its coverage of the Hayes-Tilden presidential election to assert that any residual doubts about the intra-Mercurian planet could be put down to simple professional jealousy: " 'Vulcan may possibly exist,' said the conservative astronomers, 'but Professor So and So never saw it...' "—pure us-against-them nastiness, according to the *Times*, adding "they would hint, with sneering astronomic smiles, that too much tea sometimes plays strange pranks with the imagination."

Now, such too-smart fellows were about to receive their due, the newspaper proclaimed. Why? Because, in the wake of Weber's report, the grand old man himself, Urbain-Jean-Joseph Le Verrier, had roused himself. "The man who untied Neptune with his nose—so to speak—cannot be accused of confounding accidental flies with actual planets. When he firmly asserts that he has not only discovered Vulcan, but has calculated its elements, and arranged a transit especially for its exhibition to routing astronomers..." the *Times* wrote, "there is an end of all discussion. Vulcan exists..."

The *Times* got at least one thing right. After shifting his attention to other problems for a few years, Le Verrier had indeed returned to the contemplation of Vulcan. Wolf's news had fired his passion for the planet, and he began a comprehensive reexamination of everything that might bear upon its existence. Starting with yet another catalogue of claimed sightings dating back to 1820, he identified five observations spread from 1802 to 1862 that seemed to him most likely to represent repeat glimpses of a single planet. That allowed him to construct a new theory for the planet, complete with the prediction the *Times* had rated so high: a transit that could perhaps be observed, Le Verrier suggested, on October 2nd or 3rd.

The headline writers would be disappointed. Vulcan did not cross the face of the sun in early October. More confounding, Weber's revelation from China was debunked: two photographs made at the Greenwich Observatory clearly revealed his "Vulcan" to be just another sunspot. *Scientific American* called this the *"coup de grace"* for this latest "discovery," but, as usual in the annals of Vulcan, its real impact was more deflating than destructive. Le Verrier's calculation turned on earlier observations, not Weber's, and there was a way to explain away the missed transit, by positing an orbit for Vulcan that was much more steeply inclined than previously assumed. Thus Le Verrier hedged his bets: there *might* be a chance to see Vulcan against the face of the sun in the spring of 1877, but given the full range of possible orbits this insufferably errant planet might occupy, it might be five years or more before the next transit would occur.

No transits occurred that March. Le Verrier said nothing more in public about Vulcan. He had turned sixty-six on March 11, and he was tired to the bone. As the year advanced, he found he couldn't drag himself to the weekly meetings of the Académie, nor to his daily post at the Observatory. Time off seemed to help—he returned to his desk in August—but fatigue masked his real trouble: liver cancer.

On the evidence, Le Verrier was not a religious man. He did accept communion in late June on the urging of a much more committed Catholic colleague, but that seems to have been the limit of his willingness to acknowledge conventional pieties. By summer's end, he could no longer mistake his illness. The end came on September 23rd—forty-one years to the day since young Johann Gottfried Galle had sought and found Neptune in the night sky above Berlin.

Le Verrier left the solar system larger than he found it—one both better and less completely understood. Of Vulcan itself, though—surely, given all the fully satisfactory explanations for the behavior of every other astronomical object derived from the Newtonian synthesis, the fault, it seemed so nearly certain, must lie not in the stars, but in some human failure to crack this one particular mystery.

7

"SO LONG ELUDING THE HUNTERS"

July 24, 1878, Rawlins, Wyoming.

The man from New Jersey had heard stories about the mythical West, but this was his first chance to compare the tales with the real thing—at least, as much of it as he could now observe it from the comfort of a rail carriage. His trip to Rawlins, on the Union Pacific line that ran to and through the little town, had thus far revealed only a sightseer's version of the frontier. "The country at that time was rather wild," he wrote. "Game was in great abundance and could be seen all day long from the car window, especially antelope."

It all got a little more up-close-and-personal at the hotel. He and his roommate had settled down for the night when "A thundering knock on the door awakened us; on opening the door a tall handsome man in western style entered the room." That visitor—on inspection, not entirely sober—introduced himself as Texas Jack. The hotel owner arrived and tried to persuade Jack to keep it down. He was bounced across the hall for his pains. Calmly, Jack "explained he was the top pistol shot of the West … and then suddenly pointing to a weather vane on the freight depot, pulled out a Colt, revolved and fired through the window, hitting the vane."

Other guests swarmed into the room to see who'd just been killed. With no corpse in sight, calm, of a sort, soon returned, and it became clear that the Texan just wanted to talk. At last, by promising to make time for him in the morning, Jack was persuaded to pack it in, and the two strangers returned to their beds.

But not to sleep. They left this first true encounter with the West of legend "rather scared," and, understandably, unsure "what would be the result of this interview." Nothing that had followed reassured them, to the point that they became "so nervous we did not sleep any that night." The next morning the travelers were relieved to find out that around town Texas Jack "was not one of the 'badmen' of whom they had a good supply." Thus reassured, the traveler from New Jersey could focus on the business that had brought him west. The great eclipse of 1878 would be coming to Rawlins in five days' time, and a tribe of visiting scientists were already racing to prepare. Among them, there to test his latest invention, came the man Texas Jack couldn't wait to meet: Mr. Thomas Edison.

The eclipse of July 29, 1878, followed a route that stretched from Siberia across the Bering Strait into Alaska, then down through western Canada. From there it traversed the United States through the northern Rockies across Wyoming, before heading southeast into the Gulf of Mexico, finally coming to an end just east and south of Haiti. The total phase of the eclipse—the time the moon completely blocks the surface of the sun, thus revealing both the wispy, achingly beautiful corona *and* any faint celestial bodies close to the solar limb—had a maximum duration of about three minutes, eleven seconds, achieved in Siberia.* In

* This isn't terribly impressive as such events go: given all the variables, the relative sizes of the sun and the moon and the variations in both the earth-moon system's orbit around the sun and the moon's around Earth, the longest possible eclipse lasts about seven and a half minutes.

Rawlins, totality would extend just two minutes and fifty-six seconds, but the site had one critical advantage: the decade-old transcontinental railroad ran neatly along the eclipse route, which meant astronomers and their bulky, awkward equipment could ride in unprecedented luxury to (hopefully) perfect vantage points.

Luxury is perhaps too strong a word. Wyoming had become a US territory only ten years before—the arrival of the railroad was no coincidence—and it was still very much a frontier outpost.* The Great Sioux War—during which the Battle of the Little Bighorn was fought—had ended only the year before, while in 1879, troops stationed at Wyoming's Fort Steele would move against Ute bands protesting repeated white incursions onto their land. Fort Steele's commander died in that campaign. The summer before he had accompanied Edison on his post-eclipse hunting vacation.

In other words, Edison had some reason to be a bit jumpy. For him along with everyone else coming from the settled East, the raw, desiccated ground around Rawlins lay at the very edge of the map—and yet for a week or so, it became a mecca for American astronomical research. The federal government had funded eight posts from the Wyoming territory through Colorado and down

---

Eclipses won't be visible forever. The tidal dynamics affecting the earth and moon increase the distance between the earth and the moon—very slowly, by 2.2 centimeters, or less than an inch per year. In approximately 1.4 billion years, the moon will have drifted far enough away so that its apparent size will be too small to block out the sun.

For a very different kind of take on the fact that the earth and moon are drifting apart, see Italo Calvino's *Cosmicomics*.

* Also noteworthy: Wyoming, incorporated in 1868, became in 1869 the first US territory to grant the vote to women.

to Texas to enable scientists to track the eclipse. Several research teams concentrated on the Rawlins area, following the same logic that brought Edison: the adequately long totality there could be viewed within range of a transportation system that could carry the full arsenal of modern observing tools. And among the most prized trophies that brought them all to Rawlins? The outstanding solar system mystery: where, if anywhere, the elusive Vulcan might be seen.

Henry Draper, a physician-turned-astronomer and a pioneering astrophotographer, led the largest expedition in town. Edison joined Draper's party to pursue a technical goal of his own, testing a device he called a tasimeter, an infrared measuring instrument so sensitive that he wanted to see if it could detect faint IR radiation from the corona. Along with him came Norman Lockyer, probably the best-known scientist in the group. Founder of the journal *Nature*, he was one of the pioneers of the new technique of spectroscopy. In 1868 he had noticed a bright yellow band in the spectrum of solar light, which led him to identify the element helium—the first to be found beyond Earth, untouched by human hands. Then there was the man on a mission: James Craig Watson. Director of the Ann Arbor Observatory, Watson was the veteran of two prior eclipses and had discovered more than twenty asteroids. His reason for being in Wyoming was simple: Vulcan. The few minutes of daytime darkness during totality would be, as every astronomer knew, the perfect time to detect any intra-Mercurian bodies.

He had company—or competition. Simon Newcomb from the Naval Observatory in Washington, still making his reputation as the preeminent analyst of the solar system to follow Le Verrier, also had his sights set on Vulcan. He had planned to set

Eclipse hunters at Rawlins, Wyoming, July 1878. Thomas Edison is second and James Craig Watson is sixth from the right.

up his viewing station at the town of Creston, about thirty miles west of Rawlins, but when his advance men tested their site, they found that "at the point selected a violent westerly gale was blowing without any ready means of securing the instruments against its force." The problem wasn't just the sheer intensity of the wind. In the rain shadow of the Rockies, the long Wyoming slope was high desert country. Even a moderate blow kicked up a skein of dust thick enough to turn an eclipse into a shadow play.

Newcomb's scouts turned east along the Union Pacific right-of-way, heading toward Great Divide Basin, just west of Rawlins. The railroad had placed outposts every few miles along the slow climb from Cheyenne and Laramie toward the continental crest. One, about equidistant between Rawlins and Creston, was a fly-speck of a place. The label on the Union Pacific maps identified it as Separation, Wyoming.

At its height, Separation never amounted to more than a tele-graph office, a couple of rough houses, and a water tower. Today, to find where it used to stand, you have to look just to the south of Interstate 80, about thirteen miles past Rawlins. There's noth-ing there now, no way to tell that human beings had once man-aged to scratch out a settlement. But in 1878, that's where Newcomb's men found "a small plain about fifty yards in extent, depressed below the general level and flanked on the south and west by a nearly perpendicular natural parapet some ten feet in height." Simon Newcomb joined his heralds a few days later. In all, his expedition mounted four telescopes, one set up for astro-photography, and another, along with two chronometers, desig-nated specifically for the search for Vulcan.

July 21, 1878.

The weather was a constant worry for the eclipse observers; it always is. As the days passed at Separation, a pattern developed: clear mornings in southern Wyoming with clouds stacking up in the afternoon—when the eclipse would occur. There was some hope, as the cloud cover appeared later and later each day, but there was no telling what would happen on the 29th.

Besides the weather, eclipse observers suffer nightmares thinking about the sudden-death nature of their science. Astro-nomical measurements are hard enough to make in controlled laboratory settings or within established observatories. The re-searchers at Separation sought to capture faint, small, highly un-certain observations with delicate and complicated equipment set up on uneven desert ground at an altitude of roughly 7,000 feet (2,150 meters)—with less than three minutes to get everything right. As the day of the eclipse approached, the tension within Draper's party and Newcomb's would have become as tight as a tie-down braced against a Wyoming summer wind.

July 29th, dawn.

The most famous description of that morning comes from the Cheyenne *Daily Sun*, which reported (in what reads now with a nasty edge) that the sky was as "slick and clean as a Cheyenne free-lunch table." Léopold Trouvelot had set up his instruments in the dug-out remains of an abandoned settler shack a few miles west of Separation and confirmed that report. At dawn on the 29th: "the sun rose clear and bright above the distant horizon of the great alkali plain; not a cloud was to be seen in the deep-blue sky stretching above us in all its purity."

Such glory did not last. By 8 A.M. as Trouvelot and his team were eating breakfast, they "found ourselves and the dishes completely covered with sand and dust, which had been forced by the violence of the wind through every opening and fissure." The astronomers at Separation chewed the same dirt. Newcomb reported that "in the forenoon the most violent gale we had yet experienced began to blow from the west and increased in intensity until nearly the time of the eclipse." Dust swiftly shrouded the sky, producing what he called "this obnoxious halo" around the sun. By noon, it became clear that the sand-ledge the astronomers had counted on to shelter their instruments couldn't stand up against the strengthening gusts, so the team scrambled, drafting soldiers detached from Fort Steele to erect sections of railroad snow-fencing. The desperation move worked—barely: the fence "required their [the soldiers'] constant attention, and even then a portion of it was blown down."

While fighting that skirmish, the Separation party found itself playing host to two new observers who had come up the line from Rawlins in a special car. The Englishman Lockyer had decided to get out of town, as had Newcomb's fellow Vulcan hunter Professor Watson. Whatever pressure he may have felt, Newcomb managed to be gracious in the face of invasion. He invited Watson to set up near his own telescope, a sensible as well as a generous move. If either man were to catch sight of an intra-Mercurian planet, the other would be right at hand to check.

Or so they hoped. Eclipses are utterly unforgiving. There is usually only one chance to perform each action; the more complicated or delicate the operation planned, the more opportunities for a fatal error. Newcomb's own telescope was the first to fall: ten minutes before the start of the eclipse, its clock drive failed "in such a way that it could not be used without taking the clock work all to pieces"—that is to say, with the finality that always seems to attend eclipse astronomy. First contact (the instant the face of the moon touches the edge of the sun) arrived remorselessly at 2:03:16.4 P.M. Newcomb gave up on his mechanical aid and tracked his telescope by hand—a task he made yet more difficult by his choice of a too-powerful eyepiece, one with such a tiny field of view that it made it very hard for Newcomb to be sure just when the eclipse truly began.

2:45 P.M. Totality minus twenty-eight minutes and some seconds.

Eclipses dance to a jagged rhythm. First contact shoots a jolt of adrenaline through any witness. What comes next soon becomes kind of dull. It takes about an hour for the moon to reach second contact—the onset of totality. For much of that interval,

HARPER'S WEEKLY

A JOURNAL OF CIVILIZATION

Vol. XXII.—No. 1130.] NEW YORK, SATURDAY, AUGUST 24, 1878.

*This **Harper's** Weekly cover is a remarkably fine
depiction of the strangeness of an eclipse.*

changes are subtle. Half a sun illuminates the world pretty much
as well as the whole disk. Slowly some surreality takes hold. For
example, during the partial phase a tree's crown becomes a camera obscura: the gaps between leaves transmit an image of the
crescent sun, bright curves dappling the shadows.

For the most part, though, the real surprise of the first half
hour or so of an eclipse is how ordinary the world still seems—
until you peer at the sun and confront that eerie curve of dead-

black slicing across its face.* The persistence of the ordinary slips as totality approaches. Perhaps most deranging, colors shift, then drain from the landscape. There are none of the cues of a sunset. Rather, the effect of pulling sunlight from the sky in the fullness of the day is just odd enough to make it seem as if reality itself has cracked. As each second ticks toward totality, the effects intensify; one feels an eclipse as much as one views it.

Eclipse veterans learn to guard against such distractions. At about a quarter to three, Simon Newcomb ducked into his make-shift darkroom to check on the photographic side of his planned work. He remained inside until three minutes before totality. At about 3:10, he emerged, his eyes adjusted to a sky grown strange. He took up station next to James Watson, already standing to his telescope.

One of the other men there kept time, beating the seconds and singing out the stations of the clock—each contact and the onset of totality. Watson, like all those with him, would have re-hearsed his plan. He was a cautious observer, and aimed to avoid an excess of ambition: he would study only a narrow strip of darkness near the limb of sun, as "from my previous experience in work of this character, I had determined not to undertake to sweep too much space." He had memorized the stars within that region of the sky, but still kept a star chart by him during the eclipse in a belt-and-suspenders move to avoid mistaking the fa-

* Never, never, never look at a partial eclipse with the naked eye or through a tele-scope or binoculars: eye damage up to and including blindness will result. See NASA's guide to viewing an eclipse safely: http://eclipse.gsfc.nasa.gov/SEhelp/safety.html, and in more detail this: http://eclipse.gsfc.nasa.gov/SEhelp/safety2.html. A nice do-it-yourself guide to eclipse watching can be found here: http://www.exploratorium.edu/eclipse/how.html.

miliar for a discovery. If Vulcan were there to be found, he had done what he could to bring it home.

3:13:34.2 P.M. Totality.

As soon as the shout came, James Watson fixed the sun in the center of his field of view. From there, he slowly swept due east. At the limit of his predefined search pattern, he moved his telescope one degree down and reversed direction, covering about eight degrees of sky each way. On his first pass, he recognized a familiar star, Delta Cancri. Back on the sun, he repeated the move heading west. Theta Cancri, another star in the constellation Cancer, slid into his eyepiece. There, so early in his run, Watson saw something new. He wrote that between the known star and the sun, "and a little south, I saw a ruddy star whose magnitude I estimated to be 4 1/2." It was definitely brighter than Theta Cancri, Watson added, "and it did not exhibit any elongation, such as might be expected if it were a comet in that position."

That star was not on his chart. A new object. No tail. Not, then, a comet. The stranger was running out of things to be . . .

Watson had come to Wyoming with a homemade device—a set of concentric disks faced with cardboard—to mark the location of any mysterious objects he might find. He set down unknown object "a," noted the time, and returned to his eyepiece. He dropped down one degree, and swept out west a second time. Another strange star appeared, but with at least two minutes gone since second contact. Watson had a choice: to look for known stars to serve as landmarks—or to mark the observation on his rough-and-ready record. Seconds ticked by. Watson scribbled the location of his second unknown, dubbing it "b."

A few yards away, Newcomb had spent the first minute or so

Simon Newcomb's drawing of the solar corona at the 1878 eclipse.

of his eclipse on the solar corona—the wispy detail that to the naked eye extends as many as ten solar diameters across the sky. He saw bright rays shining through the fainter background glow, and he paused to jot some notes on the feature. He then jumped to his second instrument, the one tasked for Vulcan. He had no illusions about the obstacle he faced: "The sky was so bright that it would be very easy to overlook a faint object unless the eye looked directly at it." His first survey found nothing more than two familiar pairs of stars, each clearly marked on his chart. Further sweeps turned up more spots of light, "but nothing that was not already on the map." As totality approached its end, he simply gambled, making "wider sweeps much at random, with the object of picking up some object by chance." One appeared at just about the last possible moment. As the final moments of totality flowed past, he held his telescope on it to fix its position.

3:16:24.2 P.M. Third contact; totality ends.

When the moon's disk slides past the face of the sun, the world jumps.

It feels unfair, as if one's been granted a moment's glimpse of an utterly different reality—that rectangle of Narnian forest through the open doors of a wardrobe, or a sudden vision of the train on Platform Nine and a Three-Quarters. Then, a crescent of sunlight appears and the normal, increasingly day-lit world returns. The corona switches off, and any stars that had appeared during totality rapidly fade. As sunlight returned to the high plain by Separation, Watson ran out of time. He hadn't managed to find any landmark stars for "b." Now, in a hail-mary bid to get at least a hint of replication, he ran over to Newcomb "in the hopes that he might, before the sunlight became too bright, get a place of the strange star I had first observed"—"a," near Theta Cancri.

Newcomb couldn't. He was still making sure of the position of the object he'd found on his last wide sweep north of the sun. Watson dashed back to his own instrument. No good: Watson could no longer make out either of his candidate objects. Newcomb later confirmed that his candidate was just a familiar star, adding, "it is of course now a matter of great regret that I did not let my own object go and point on Professor Watson's."

Watson did not seem to mind. Even without Newcomb, he had no doubt about "a": "In the case of the star observed near Delta Cancri I was sure, and the discovery was accordingly announced by telegraph." Or, as the *Laramie Weekly Sentinel* put it with some added exuberance: "Professor Watson of Ann Arbor, Michigan... had taken the job of FINDING VULCAN," and then, on a historic Wyoming afternoon, "He found it," adding, "It has come to be well understood among astronomers that Watson has a corner on the discovery of comets, asteroids, planets etc."

Vulcan! Two decades after Le Verrier had, for the second time, conjured a planet at his desk, there it was: a small ruddy object, moderately bright, orbiting the sun undeniably inside the orbit of Mercury. Watson's discovery was amplified and seemingly confirmed by a second sighting by Lewis Swift, a well-regarded amateur observer who watched the eclipse near Denver. The news rocketed around the world. Lockyer, with his front-row seat at Separation, wired the news to both the French and British national observatories. The British press picked up on the story, while *The New York Times* barely managed a scrap of journalistic restraint. Its first article, on July 30, merely noted that "Prof. Watson discovered an extra mercurial [*sic*] planet of the size of a 4 1/2 magnitude star ..." On August 8 it published Watson's claim for his Vulcan that "its position in reference to the sun and a neighboring star I determined by a method which obviates the possibility of error," which is why, he wrote, "I feel warranted in announcing it as an interior planet." On August 16, the paper followed up with a longer analysis of both the sighting and its significance, writing "One brilliant discovery will probably date from this occasion.... The planet Vulcan, after so long eluding the hunters, showing them from time to time only uncertain traces and signs, appears at last to have been fair run down and captured." The paper conceded that at least one more confirmed sighting was needed to support Watson's and Swift's, but still, the anonymous reporter's eagerness was obvious. The discovery would likely, the paper reported, "hold a conspicuous place in the annals of science."

There was, to be sure, some need for caution, and the *Times* reporter was honest enough to acknowledge it: "The negative results of Profs. Newcomb, Wheeler, Holden and others, who, with similar instruments, went over the same ground and found nothing, are, indeed, unsatisfactory and puzzling." But having allowed that caveat, the *Times* swiftly regained its good cheer: the missed sightings "can hardly outweigh the positive evidence on the other side ..." Such confidence may have made for the better headline, but the article missed its own point. Almost everyone looking at the sky in Wyoming and along the long American swath of totality failed to see what had appeared so obvious to Watson and Swift. Whom to believe?

Almost immediately, the argument boiled down to the same one that followed every purported Vulcan sighting. Was Watson's Vulcan a new planet or just a mistake, an ordinary object in disguise? Watson never confessed to any doubt: "there is no uncertainty in the place of (a)" and he was sure that "I saw both it and Theta Cancri." That almost all other eclipse observers hadn't seen it bothered him not a bit, and for a pretty good reason: any experienced worker, familiar with high-powered telescopes, he wrote, would "know how uncertain a search would be under the circumstances." True enough. But even so, veterans of the Vulcan quest had heard this before. One man gets to see it; everyone else doesn't. Again.

In the beginning, most of Watson's colleagues were prepared to suspend judgment. He was a professional observer and a good one, so the astronomical community was unwilling simply to dismiss his findings. But the sheer weight of all that nothing seen in the other telescopes pointed at the sun at the same time left most astronomers unsure—at best—about what to make of this latest vision of the elusive planet.

A few flatly refused to defer to reputation. C. F. H. Peters—discoverer of forty-eight asteroids and one of Watson's avowed rivals in the minor planet game—published a devastating attack on this latest Vulcan candidate. He accused Watson of making a series of elementary errors. Peters questioned Watson's make-shift system for marking positions. He argued that Watson couldn't have made a reliable assessment of the brightness of "*a*." He offered an explanation for Watson's description of the unusual ruddy color of the proposed planet. None of the steps in Watson's procedure survived scrutiny, leading Peters to his blunt, almost brutal conclusion: "It is, therefore quite apparent to every unbiassed mind that Watson observed Theta and Delta Cancri, nothing else."

Peters's report bubbled with scorn, and Watson reacted with a mixture of formal defense and outrage, writing, "Professor Peters's whole attack upon the integrity of my observations is not of the slightest consequence, since he has created the errors in his own brain," adding "I do not intend to engage in any controversy about these matters and especially with a person who was, at the time of the observations, more than two thousand miles away." Publicly, his colleagues gave him tepid support. Commentary in the journal *Nature* chided Peters for his tone, writing "Throughout Prof. Peters's criticisms ... there is evinced a certain *animus* which could have been as well avoided." Fair enough, but for all its seemingly impartial account of the Peters-Watson fight, the *Nature* piece subtly made its judgment clear. "We will venture to say that the general feeling amongst astronomers when first reading Prof. Watson's announcement ... [is that he] would not risk his whole scientific reputation by putting forth such a statement to the world, unless he was firmly convinced of its truth." The hammer fell in the next few words: "Otherwise, the fact that there were two known stars on the parallel or nearly so and less than one degree west of the objects supposed to be new, would probably have been felt to be an almost fatal objection to the reality of the discovery."

Such delicate language muffled the blow, but the point was plain enough. Peters may have been a rude SOB, but he had said out loud what was rapidly becoming the consensus of astronomical opinion: "a" and "b" were nothing more than "two known stars," misidentified amid the furor of an eclipse, recklessly trumpeted as discoveries in the adrenaline of the moment and preserved in the heat of desire, the felt urgency to make real what must be there. Watson disagreed, of course. He never retracted his Vulcan in what turned out to be the brief time left to him. In the autumn of 1880, he came down with a sudden infection and on November 23, he died. He was forty-two.

With Watson silenced, many, and not just the vengeful Peters, felt liberated to say out loud what they had previously only whispered. The old pattern held: Vulcan was to be found only when it was sought in the faith that it had to exist, never when anyone tried to confirm what had persuaded someone else. The astronomical community's consensus quickly hardened. James Watson had seen what he longed to see. His "Vulcan" was nothing more than a mistake.

Almost a full two decades earlier, Vulcan had gained entry into the solar system. The eclipse of 1878 signaled its banishment. By the 1880s, the old saw flipped: absence of evidence had finally accumulated to the point where (to almost everyone) it had indeed become evidence of absence.

July 29, 1878, evening.

Thomas Edison knew his experiment had failed almost immediately. His tasimeter was not sensitive enough to pick up infrared radiation from the solar corona. That null result couldn't dent his good mood, though. This western trip, he had told reporters, was the first vacation he'd taken in sixteen years, and he was prepared to enjoy himself, no matter what.

For their part, his hosts were eager to entertain their famous visitor, but no one let Edison suffer any delusions about his status: he was a tenderfoot. Edison got a taste of western humor a day or so after the eclipse, when he and some companions took a rail excursion up to Separation. Edison packed along his Winchester rifle, on the chance he might bag some local fauna. At the depot, the tourists were greeted by the station agent, John Jackson Clarke. Clarke wasn't terribly impressed with the outdoor skills of his visitors—"their combined knowledge of game killing," he wrote, "was about equal to mine of parallaxes and spectrums." Out went the intrepid hunters anyway, straggling back that evening having bagged between them a grand total of exactly one sparrow hawk.

Edison returned to the station first, and he asked whether there might be anything else worth shooting nearby. Clarke told him that the surrounding plain enjoyed an abundance of jackrabbits—"what the locals call narrow-gauge mules." Edison asked where he might find them, and Clarke "pointed west and noticing a rabbit in a clear space in the bushes, said there is one now."

Edison picked out a silhouette from the platform, but he wanted to make sure of his kill. He "advanced cautiously to within 150 feet and shot."

The animal did not move. He closed to one hundred feet. He fired again.

The beast wouldn't jump. He aimed, pulled the trigger once, and then again.

His target stood its ground.

Edison glanced over his shoulder and saw that the entire station staff had gathered for the show.

The penny dropped.

He'd been set up, played for a dude. His target looked like a desert hare, all right, all ears and legs. It was exactly where one might expect to spy such an exotic creature. And yet...

Thomas Edison, genius, had just murdered ... a stuffed jackrabbit.

It had seemed so real.

"A SPECIAL WAY OF FINDING THINGS OUT"

Vulcan after the eclipse of 1878 became—to steal a twentieth-century trope—something of Schroedinger's planet. Like the famous cat, as long as no one actually looked for it, an intra-Mercurian mass made such perfect sense that it gained a kind of *potential* existence. It was there/not there; in and out of sight; logically necessary, yet absent.

The confounding issue remained, of course. Mercury still misbehaved. Simon Newcomb was the most authoritative student of the solar system in the last years of the nineteenth century. In 1882 he redid Le Verrier's calculation and showed that Mercury's excess perihelion advance was slightly larger than Le Verrier had originally determined. But the drama of the Wyoming eclipse left astronomers with few choices. Vulcan, whether imagined as a single planet or a flock of asteroids, was no longer plausible as the source of Mercury's anomaly. What to do?

No-shows are hardly alien to science.

Theories predict. That's their job. Ever since Newton and his co-conspirators consummated their revolutionary program of subjecting nature to mathematics, this has come to mean that particular solutions to systems of equations can be interpreted as physical phenomena. If a given mathematical representation hasn't yet matched up with some phenomenon in the real world, that's what's called a prediction. From the theory of Uranus, Neptune; from the theory of Mercury . . . what, if not Vulcan?

But what happens when a prediction fails to find its match in nature? This is a constant question in science. Take one recent example: for half a century, there was the mystery of something called the Higgs boson. The Higgs is the quantum, or the smallest possible change in energy, in what is known as the Higgs field. The Higgs concept was first proposed in the mid-1960s as part of what is now called the Standard Model of particle physics, a theory that describes the properties of the elementary particles out of which reality is built.* Within the Standard Model, the Higgs boson accounts for how certain of those particles acquired the mass that they have in fact been seen to possess.

Over the next several decades the Standard Model proved phenomenally successful, its predictions matching experimental results to as many decimal places as any measurement could achieve. But not the Higgs—which stubbornly refused to appear.

The Higgs was finally captured in in observations made in 2012 and 2013, following the construction of the Large Hadron Collider (LHC), an instrument powerful enough to peer into domains invisible to earlier devices. Up until the machine produced its data it remained an utterly open question as to whether the Higgs would actually show itself at the energies the machine could produce.

And if the LHC hadn't found its Higgs? That would have been

---

* To call something a fundamental particle is to say that it is a chunk of reality that cannot be broken up into any smaller components. Another example of such a particle would be the photon —the quantum, or finest possible subdivision of light across the entire spectrum of electromagnetic energy.

a direct analogy to the problem Vulcan after 1878 seemed to pose (for all that no one addressed it): the failure to find the result theory anticipated in a context that demanded *some* solution would raise deep and (for theoretical physicists) very exciting questions.

The Higgs is no isolated example. Take, for example, the mysteries that remain in the account of what happened as our universe was born. So much has been discovered about that seemingly inaccessible time and process because the Big Bang—the explosive appearance of space and time, matter and energy, essentially out of nothing*—left a snapshot of itself in a flash of light called the Cosmic Microwave Background, or CMB. Discovered in 1964 as a seemingly uniform hiss of microwaves (the same year that the Higgs idea first emerged), the CMB offered the chance to do something new: to measure detailed properties of the very early universe by extrapolating backward from that microwave glow to the Big Bang process itself.

In the decades since, the interplay of cosmological theory and ever more refined observations has yielded a series of insights about that nascent universe, along with predictions about what kinds of features should be found in the CMB. For example: just by looking around us, it becomes obvious that the present-day universe is lumpy, with big piles of matter collected into stars and galaxies and clusters of galaxies—and giant, mostly empty

* In current theory, a particular kind of nothing called the false vacuum. False vacuums are regions of space-time that appear to be truly empty of phenomena, but through the effects of quantum mechanics can be populated, seemingly from nowhere, by subatomic particles or energy fluctuations.

spaces in between. What we see now implies that the CMB should clump too, that there should be places in the microwave picture of the universe that shine just a little brighter than other places: hot spots that map the slightly more matter-rich neighborhoods that could ultimately grow into galaxy clusters.

Early surveys of the microwave sky, though, showed a completely uniform, blank glow. If that were all there was, such a featureless early universe would seem to be incompatible with what we know is out there now—and that in turn would imply that what cosmologists thought they knew about the cosmological evolution was wrong.

That's how matters stood for almost three decades until 1989, when a specialized telescope was launched into Earth orbit. By 1993, that instrument had captured enough photons to reveal exactly a broad pattern of light and dark—the first, out-of-focus glimpse of the original "seeds" of galaxy clusters. There was a prediction based on a clearly observed fact in the contemporary universe ... and through enormous effort, it was shown to be true.

Since then, the CMB has been studied at greater and greater resolution to reveal an increasingly detailed picture of the events that turned the infant cosmos into one recognizably like our own. At the same time, theorists have made a series of predictions to be tested when and if observations of the CMB could be improved yet more. One idea first proposed in the 1980s suggests that during its first instants of existence, our universe underwent an episode called *inflation,* during which space itself expanded at a ferocious rate—the bang of the Big Bang itself, as one of its inventors, Alan Guth, describes it. For more than thirty years, observations have yielded results that are consistent with inflation, but despite that growing body of evidence, open questions remained.

That seemed set to change in 2014, as researchers closed in on a key expectation of the theory: that inflation's wild ride would create what are called gravity waves, ripples in the gravitational field that would show themselves in particular (and very subtle) features that might be detectable in the CMB. There are several versions of the idea, each of which predict somewhat different signals. In some of them, those primordial gravitational waves would leave a specific imprint on the CMB as a particular type of polarization within the microwave background—thus revealing the first unequivocal connection between the vast, fast madness of the inflationary universe with our own, more sedate cosmos. If such effects were found, it would be the final rung in the ladder of observations—the smoking gun to confirm that we really do live in an inflated universe.

That was the mission a research team set for itself with its instrument at the South Pole. The BICEP2 microwave telescope started gathering polarization data in 2010. The team ran it for two years before beginning to study its data in earnest. It was a delicate, difficult analysis, and the stakes in the answer were so high that the researchers took every precaution they could think of to make sure they got it right. The public announcement came on March 17, 2014: B-mode polarization had been observed in the CMB to a 5.9-sigma, which was much better than the 3.5 million–to-one level of certainty required to claim discovery.

It was a thrilling moment. The result made front pages around the world. It brought one of inflation's inventors to tears. For scientists and amateurs of science alike it was a gift: something beautiful, strange, and newly intelligible about existence on the largest scale. There was a distant resonance, an echo of what those first few must have felt in 1687, when the earliest copies of *Principia* came into their hands: a kind of breathlessness, sheer wonder that human minds could penetrate such incredibly deep mysteries. One of the most persuasive readings of inflation is that we dwell not in a singular cosmos, but in just one of un-counted island universes, our little village within a vast Multi-verse. What a thought! No wonder that a grown man wept to hear the good news.

Observing at the ragged edge of technology is always a tricky business. Each Vulcan "discovery" drew informed scrutiny. So did the tiny fluctuations the BICEP team found within their data, the signal they claimed was the signature of inflation's gravity waves. Questions about their results became doubts within a few weeks, as scientists from outside the team pressed on the issue of foreground dust—perfectly ordinary debris common in galaxies like our own Milky Way. By summer's end, it had become clear that the filtering of light through such nearby dust might explain all of the effects visible in BICEP data. Planet or sunspot? Multi-verse or stellar schmutz?

Mercury still precesses and by many measures the universe behaves *as if* inflationary theory is correct, but by early 2015, at-tempts to check the BICEP2 measurement confirmed that it was impossible to distinguish a clear answer, given the confounding role of the galactic dust. As in 1878, the mystery remains open; we still do not know what happened during the birth of the par-ticular cosmos we inhabit. There is one important difference, though, between those searching now for ripples in space-time and those who after the eclipse of 1878 gave up on Vulcan. What is known to date is that BICEP2 results do not contain a reliable observation of inflation's signature in the CMB. That doesn't (yet) mean such traces don't exist. Several attempts are already under way to probe the CMB with yet more precision. Those measurements will likely settle whether the predicted gravity waves really do reveal themselves in the microwave background, and even if the hoped-for polarization effects are not found, there are versions of inflation theory that do not require a gravity wave signature in the ancient glow of the Big Bang.

Still, even if some form of inflation remains a persuasive candidate to account for the properties we see in the universe right now, it hasn't closed the deal. The cosmos could see things differently.

Long gaps between prediction and observation always raise the question: what finally persuades science—scientists—to abandon a once successful idea? When do you take "no" for an answer? There's a conventional response in science to that question: right away. Or at least as soon as you're confident of the evidence. In a public talk delivered in 1963, Richard Feynman said that science is simply "a special method of finding things out." But what makes it special? The way its answers get confirmed or denied: "Observation is the judge"—the only judge, as the catechism goes—"of whether something is so or not."

There is a strange magic to the term "the scientific method." At a minimum, it asserts a particular kind of authority: here is a systematic approach, a set of rules, that when followed will reliably advance our understanding of the material world. Such knowledge, though, is always provisional, a seeming weakness that is the real strength of science: every idea, every generalization, every assumption is subject to question, to challenge, to refutation.

That's how the scientific method is usually taught. Every high school student confronts some version of Feynman's description. The process of science rides down railroad tracks: you "Construct a Hypothesis" to "Test with an Experiment" (or an observation), and then you "Analyze Results" and "Draw Conclusions." If the results fail to support the initial hypothesis, then it's back to step one.

Laid out like that, the scientific method can be seen as a kind of intellectual extruder. Set the dials with the right question, pour data into the funnel, and pluck knowledge from the other end. And, most important: when that outcome fails to match reality, then you go back to the beginning, work the dials into some new configuration, try again.

This isn't just cartoon stuff either, a caricature told to children who may never dive more deeply into science than a Coke-Mentos volcano. Even for those who penetrate into more and more advanced ideas and approaches, the same message gets dressed up in more formal language. Here's a typical "Introduction to the Scientific Method" aimed at college students: "The scientific method requires that a hypothesis be ruled out or modified if its predictions are clearly and repeatedly incompatible with experimental tests ..."—pretty much exactly what science-fair contestants are told. The explanation goes on, though, to echo Feynman's point: "No matter how elegant a theory is, its predictions must agree with experimental results if we are to believe that it is a valid description of nature. In physics, as in every experimental science, 'experiment is supreme.'"

In other words: when a long-anticipated outcome fails to materialize, more than a single prediction lies in peril. If gravity waves don't show up in ever more acute CMB measurements, then at some point the strand of inflation theory that requires them will be in trouble. Along those lines, once Vulcan refused to appear, decade after decade, what should have been done about that icon of the scientific revolution, Isaac Newton's theory of gravity?

Within the myth of the scientific method, there should have been no choice about the next move. "Experiment is supreme" ... "Observation is the judge." We hold this truth to be self-evident: the hard test of nature trumps even the most beloved, battle-tested, long-standing idea.

Does history behave like that? Do human beings?

No. Real life and cherished fables routinely diverge.

After July 1878, almost all of the astronomical community abandoned the idea that a planet or planets of any appreciable size existed between the Sun and Mercury. But that broad consensus did not lead to any radical reassessment of Newtonian gravitation.

Instead, a few researchers tried to salvage the core of the idea with ad hoc explanations for Mercury's motion. The historian of science N. T. Roseveare has catalogued the struggle, dividing it into two main strands. Simon Newcomb followed his recalculation of Mercury's orbit with a review of the "matter" alternatives—Vulcan-like explanations that depended on coming up with a source of mass that for some good reason remained undetected but could generate enough gravitational tug to produce the perihelion advance. He took Vulcan itself as clearly refuted, but he catalogued a number of more subtle suggestions: perhaps the sun was sufficiently oblate—fat around its middle—that such an unequal distribution of matter could solve the problem. Alas, the record of solar observation persuaded Newcomb that our star is pretty nearly spherical (as it is). Other proposals—matter rings, like those around Saturn, or enough of the dust that was known to exist near the sun—fell to a variety of other objections. After more than a decade of thinking about the problem, Newcomb came to his uncomfortably necessary conclusion: within the framework of the inverse square law of gravity, there was no plausible trove of matter near the sun that could account for the motions of Mercury.

With that, if science as lived matched the stories scientists tell about it, Newtonian theory should have been for the chop. In the fairy-tale version of the search for knowledge, Newcomb's verdict—that there was a persistent, unrepentant anomaly current theory could not explain—would compel researchers to question its status as "a valid description of nature."

In any myth there's at least a hint of some deeper truth, and so, as matter-based ideas fell, Newton's version of gravity did come under a bit of scrutiny. One astronomer suggested that Newton's law might be only an approximation: gravity could vary by masses involved and inversely with the distance between them to a power of 2 . . . plus just a tiny amount: .0000001574. That would bring Mercury's motion into perfect agreement with the math, but there were several obvious objections. For one, it was such a messy move: why would the inverse exponent for gravity "choose" to be so close to a perfect integer, and yet refuse to settle on exactly two?

To be sure, nature sometimes just *is*, in ways that can seem both arbitrary and unlovely. Even now, there are several numbers in fundamental theories of the large and small that are set by observation. In some cases they are just as odd—or weirder still—than an inverse 2.0000001574 power law. For example, the fine-structure constant, a number involved in describing the ways charged particles (like electrons) interact, is known by observation to be $7.2973525698 \times 10^{-3}$. There is nothing in any theory that offers any explanation for why that number should take that particular value. It's just the way the universe does that particular job. To Richard Feynman, this was cosmic bad taste: "All good theoretical physicists put this number up on their wall and worry about it. . . . It's one of the *greatest* damn mysteries of physics: a magic number that comes to us with no understanding by man."

Even so, simplicity, elegance, and above all, consistency have proved to be pretty great ad hoc measures of theoretical insight, even if they give no guarantees. An inverse-not-quite-two law was ugly enough that very few researchers took it seriously. The idea finally went away in the 1890s when it was shown to account for Mercury's motion, but not that of the earth's moon.

A few more attempts to tweak Newton followed. Some added another term to the classic inverse square law to better fit theory to nature, and others explored the idea that the speed of a body might change its gravitational attraction. None gained significant support from either physicists or astronomers, and they all would collapse under a variety of fatal flaws.

By the turn of the twentieth century, most researchers had given up. There was still no explanation for Mercury's behavior—but no one seemed to care. There was so much new to think about. X-rays and radioactivity had opened up the empire of the atom. Planck's desperate creation of the quantum theory was about to transform the study of both energy and the fundamental nature of matter. The decades-in-the-making confirmation that the speed of light (in a vacuum) was truly constant was beginning to hint that extremes of speed might produce some very interesting effects. Henry Adams at the Paris Exhibition of 1900 marveled at the practical applications of the new science of electricity. In 1903, the Wright brothers' experiments on a beach in North Carolina would usher in an age in which, among much else, long-pondered and very difficult questions in physics—like the motion of air over a surface—took on literally life-and-death significance.

And through it all, good old Newtonian theory worked a treat, pretty much all the time. Its laws of motion described the experience of the real world close to perfectly, and, if Mercury acted up a little (so little! those few arcseconds per century!), comets and Jupiter and falling apples and just about everything else that could be observed proceeded on their way in calm agreement with the rules laid down in the *Principia*.

Amid all this—the tumult of the new and the excellence of the old—Vulcan itself dwindled into a mostly forgotten embarrassment, the physical sciences' crazy uncle in the attic. There it sat (or rather, didn't), hooting in the rafters. No one seemed to hear. Mercury's perihelion still moved. The gap between fact and explanation remained.

That would change—but only after a young man in Switzerland started to think about something else entirely, nothing to do with any confrontation between a planet and an idea. There was a question he'd begun to ask. One way we now reframe his problem is to ask how fast gravity travels from here to there, from the sun, say, to Earth. But that's not the way it struck him on an autumn afternoon in 1907 as he stared out his window on the top floor of the patent office in Bern.

Part Three

VULCAN TO EINSTEIN

(1905–1915)

8

"THE HAPPIEST THOUGHT"

November 1907.

Albert Einstein was always a conscientious employee. He had been a model civil servant since 1902, when the Swiss national patent office took a chance on the then-unemployed recent graduate with a bachelor's degree in physics. In 1905 he had experienced what would later be called his *annus mirabilis,* his miracle year—really just six months—when, to a much greater extent than most realize, he laid the foundations of the twentieth-century revolution in physics. Almost instantly, he was transformed: no longer a mere amateur, stealing time to calculate at his government desk, he became a full participant at the highest level of international physics. And yet, he was still a bureaucrat. In 1906 his superiors promoted him to Technical Examiner *Second* Class—undoubtedly the most famous patent clerk in Europe.

He still did his job and did it well, delivering an extremely competent day's work for a day's pay. He reviewed the documents and technical drawings that crossed his desk. He wrote his evaluations, doing his part to maintain the legal framework for invention. Even so, he couldn't keep himself from pausing every little while to think about what truly moved him. So it was one day in 1907, he found himself staring out the window. Across the way, he saw a man fixing something on a roof. His imagination took over. In his mind's eye, that suddenly luckless roofer slipped, slid, fell—and there it was, what Einstein would call "the happiest thought of my life." It had just come to him that "if a man falls freely he will not feel his own weight."

A man crashing to his death would seem to be an odd image to evoke joy in anyone. And that treacherous roof was a very long way from the limb of the sun and the realm Vulcan had been supposed to roam. Even so: there stood an anonymous laborer, unaware of the mental play going on in the office across the way and, equally unknowing, about to take on a vital role in settling the fate of an undiscovered planet.

Of course, the work to come did not emerge wholly formed from that first insight. Einstein's greatest discovery built on an earlier one from his almost ridiculously profligate outpouring in the first half of 1905, when he produced four papers spread across huge swaths of theoretical physics. The first emerged from what might seem like a surprisingly narrow investigation into a phenomenon called the photoelectric effect, originally observed in 1887. A better picture of the phenomenon came at the turn of the century, through the work of the great experimentalist (and horrible man) Philipp Lenard. Lenard studied what happened when he varied the intensity of the electromagnetic radiation—light—striking a metal target. The expected result, based on Maxwell's description of light as a field of waves, was that a bigger wave (brighter light) should impart more energy to the electrons. But Lenard found that while dimming or adding light altered the amount of current produced—the number of electrons emitted—it didn't affect the energy of each individual electron as it left a metal surface. That varied only with the color of light he chose, its frequency or wavelength. Ultraviolet radiation, for example, imparted a bigger kick than did longer wavelengths—lower frequency visible colors. Lenard, who won a Nobel Prize for his experiments, couldn't explain this fissure between theory and observation. In 1905, with no formal training beyond his undergraduate degree in physics, Einstein did.

What if, he asked, light could be understood not simply as a wave, but as a kind of particle, a quantum of light—what is now called a photon. Starting from that physical intuition, interpreting Lenard's experiments becomes simple (though not easy). If light is made up of particles, then more photons (more light) would produce a bigger flux of electrons and hence more current—as observed. But the energy imparted to each of those electrons would depend on the energy of the photon that whacked it, not the total number of particles that struck the target. Once Einstein represented light as quanta in his equations, the calculation that followed reproduced Lenard's results . . . and helped form the foundations of quantum mechanics, a set of ideas that is utterly intertwined with every facet of twenty-first-century life.*

That came in March. April brought Einstein's proof of the existence and size of atoms and molecules, an exercise in statistical

---

* It would take another book—and at least thousands have already been written on this theme—to trace the tendrils of quantum mechanics in contemporary life. From the electrical phenomena that allow my computer to turn the motion of my fingers into letters on a screen and thence a page, to the specific properties of the materials within the seat upon which I sit, to the high poetry of the theories of the cosmos with which this book is partly concerned . . . quantum ideas are implicated everywhere. Most of the way we move around our world is classical, Newtonian, accessible. The secret life that underpins that overt one is inconceivable except as expressed in the language of the quantum revolution. Here the sermon endeth.

physics that remains the most frequently cited of his 1905 works, with applications that range from mixing of paint to Einstein's own definitive explanation for why the sky is blue.

He followed that up with a related analysis that solved the long-standing mystery of Brownian motion—first observed in the random motion of dust or pollen in water. That sounds like a sidelight, a minor result, except that Einstein's method of accounting for the outrageously large number of molecular collisions required to produce the wandering track of a pollen grain was a significant step in building perhaps the single most powerful idea in twentieth- and twenty-first-century science: the recognition that the fundamental nature of reality in many of its facets is determined by the behavior of crowds that can only be understood in statistical terms, and not by direct links in a chain of cause and effect.

Einstein sent off the Brownian motion paper in the second week of May. With that he'd completed work that would have been the pride of at least two careers—and when he finally won his Nobel, the prize committee cited his account of the photoelectric effect instead of the more popularly celebrated work to come. But he had one more shot to fire. The last of Einstein's "big four" papers arrived at the offices of *Annalen der Physik* on June 30, 1905. It came under a seemingly bland title, "On the Electrodynamics of Moving Bodies"—which masked the radical, almost subversive idea it contained, what we know as the special theory of relativity. It took him about six weeks to produce it, but once he was done, he was able to express his ideas in astonishingly simple, clear language, almost a story, in which he asks his readers to question what almost no one had ever paused to consider. What does it mean, he asked, to say that an event happens at a certain time? "If," he wrote, "I say that 'the train arrives here at 7'o'clock' that means, more or less, 'the pointing of the small hand of my clock to 7 and the arrival of the train are simultaneous.' " In other words—to describe any event in nature you need a rigorous concept of time: how it is to be measured and how any two people can come to agree on when anything may be said to have happened.

From there, Einstein lays down the two pillars on which all the rest of his new idea will rest. One was the "relativity principle," originally defined by Galileo. It holds that "the laws governing the change of state of any physical system do not depend" on whether someone observes that event from within a system or from the outside, looking in—as long as both vantage points "are in uniform motion relative to each other." That is: it doesn't matter whether you are standing by the track or riding a train. Newton's laws of motion (and any other natural laws, of course) behave the same way in both circumstances, even if, say, the path of a ball thrown on the train *looks* different to people watching from either vantage point.

Einstein's second axiom was that the speed of light in a vacuum must be a constant, identical for all observers throughout the universe. The problem with that idea—and this had troubled scientists for decades before Einstein—is that if the speed of light truly does remain constant for all observers, that would seem to contradict Newton's ideas about motion. Here's the difficulty: imagine that a person turns on a lantern and stands still, while another runs down its beam of light. If Newton were right, the person at rest should find light traveling at the usual number—very nearly 300,000 kilometers (186,000 miles) per second. But the person in motion should come up with a different answer: 300,000 kilometers per second, less the speed at which she runs, say twenty kilometers an hour.* To Einstein's predecessors, that's how a properly organized universe would behave. Yet stubbornly, throughout the last years of the nineteenth century, the measured speed of light never complied, no matter how precise the experiment, no matter the state of motion of the experimental apparatus.

Einstein's insight was to take seriously the implications of that evidence of a constant velocity for light. If the speed of light does not change with the motion of an observer, he argued, then to reconcile that fact with the rest of experience requires a change in the way one must think about the elements of speed—distance and time. Another thought experiment captures what he was trying to express: imagine a train traveling at a steady pace along a straight stretch of track, with one person equipped with a clock on the train at its midpoint, and another with an identical clock standing on the embankment. Now picture two bolts of lightning striking each end of the train at the instant that the watcher on the train passed her luckless counterpart, standing by the tracks amid the storm. So here's the question: do both observers agree that the two bolts of lightning hit the train at the same time?

---

* In more detail: Maxwell's equations yield a constant velocity for an electromagnetic wave (of which visible light is simply a particular set of wavelengths). By Galilean invariance, the assertion that the laws of physics remain the same for every observer in uniform motion, that constant velocity remains a constant in every frame of reference. But as Newton developed his mechanics based in part on Galileo's work, the speed of light was no more a constant than any other velocity, and the claim of Galilean invariance did not apply; hence the conflict.

Special relativity began to form as Einstein understood that the answer has to be "no." The bystander sees the strikes as simultaneous, but the train's rider does not. How can this be, given that both of them are describing the same events? Einstein's answer is, in effect: think about how the constancy of the speed of light affects your ability to decide when something happens, when the event and the tick of your clock occur at the same time. The images of the lightning bolts from both ends of the train have to cross some distance from where they strike to where the two watchers happen to be. To reach the observer standing on the embankment, the signal—the light from each bolt—has to cover the same distance: half the length of the train. Each signal will take the same amount of time (the speed of light is a constant, fixed for both bolts) to cover an identical amount of ground— and the observer can clearly describe the event: he sees the two strikes as simultaneous.

But for the watcher on the train, the situation is different. She's still moving as the bolts hit. In the time it takes for the light to travel from the strikes to where she stands, she and the train will have traveled forward just a little. The light from the strike at the front will have a slightly shorter stretch to cross before it reaches the eye of the traveling observer than the light from the rear, chasing the advancing motion of the train. This observer will see first the flash from the forward bolt, and then after a moment, the flash from the trailing one. In other words, two strikes occur at different times for this observer, one slightly before her counterpart standing next to the train sees his "simultaneous" flashes, and one slightly after. These two people in two different states of motion cannot agree on the timing of the identical events.

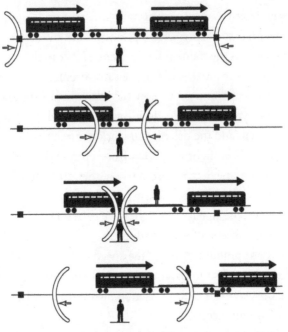

This version of Einstein's simultanaeity thought experiment depicts
what happens from the point of view of a trainspotter on the embankment. Lightning bolts strike both ends of the train as a passenger riding
at the train's midpoint passes the observer (top drawing). Einstein
wanted to know if both the observer and the passenger would agree
on the timing of the strikes. In the second drawing, the train moves
forward, shortening the distance the light from the bolt at the front
of the train has to travel to reach the passenger—which results in
that light reaching the passenger before the trainspotter. In the third
drawing, light from the forward bolt arrives at the trainspotter just as
light from the back bolt reaches him too. To him, the lightning bolts hit
simultaneously. Finally, in the bottom drawing, as the train moves yet
farther, light from the rear bolt catches up to the passenger, who thus
concludes that the two strikes hit at different times, disagreeing—in
good faith—with the observer at trackside.

What's more, Einstein realized, this disagreement holds true for distance too. Following the same reasoning as he did for time, Einstein emphasized the importance of the measuring apparatus, in this case an ordinary ruler. Imagine the passenger measuring her legroom with a ruler as she passes the train-spotter beside the track. As the rider notes her result, her measuring stick is moving past the observer on the other side of the window. The observer on the embankment measures the time it takes for the front and the rear of the measuring stick to pass him by. But as we already know through the lightning bolt thought experiment, those times will be different from the measured flow of time on the train—and from that follows that each observer, with their differing measurements of time *have* to disagree about the length of the measuring rod. Space and time are relative.

One step more remained. In his version of relativity, Galileo had worked out a set of mathematical rules—now called the Galilean transformations—to enable two people in motion relative to each other to see that their different perceptions were really equivalent accounts of the same event. Einstein's relativity established the physics within an updated set—called the Lorentz transformations—that use that single constant, the velocity of light, to reconcile any two observers' differing observations.

Newton's God kept absolute time and absolute space throughout the universe, a divine clock striking the same hours at every point throughout all creation. That article of faith helped Newton to his genuinely revolutionary insight that the heavens above and the earth below are governed by a single set of laws, just one system of the world. As the flight of comets and the discoveries of planets seemed to prove, cosmic history seemed to possess a universal constancy, the same everywhere for all people at all times. Two centuries on, Einstein's homely images of trains and time-pieces and rulers laid waste to all that. His clocks tallied their seconds beautifully, but to a beat that varied in the eye of the beholder.

The second half of Einstein's relativity paper extended its concepts beyond the realm of the motion of material objects. Relativity, he showed, held for objects in space; for the constituents of atoms; for electromagnetic fields, seemingly, for everything. Three months later, Einstein wrote again to show how deeply the concept penetrated into the fabric of what had seemed to be settled physics. In just two pages, he investigated what happens when a chunk of matter emits energy—any form of radiation for example. From that starting point, he performed a brief calculation that revealed that within relativity energy and mass can be shown to be equivalent. He wrote his conclusion in a very different form from the way we now know it, as the most famous equation in science: $E = mc^2$.

On its own the demonstration that matter and energy were really two sides of a single coin would have been a fine piece of work. Though the calculation was easy—trivial, almost—it still seems almost unbelievable that an argument about kinematics, the properties of motion, could morph into such a deep claim. Common sense says that energy is something that happens *to* matter, the strike of a bat on a ball or the explosion that throws a shell out the cannon's mouth. But common sense is wrong. Einstein's equation forces us to conceive of matter and energy as intertwined, capable of transforming, one into the other.

If that weren't strange enough, this little paper went deeper still, embodying an insight that Einstein grasped before anyone else: relativity wasn't so much a specific law of nature as a condition to which all the more ordinary patterns in nature had to conform. $E=mc^2$ translated the concept of inertia into relativistic terms. Through Einstein's further work and that of others, the laws of motion fell into line. The electromagnetic equations of Maxwell had to be reinterpreted to account for the relativity of space and time...and so on. In the metaphor of the day, relativity was an imperialist, colonizing ever greater tracts of physics. The logic of empire is to grow, and the special theory's next target was, if not obvious, then inevitable.

Einstein would remain at the patent office into 1909—among other virtues, it paid better than entry-level academic jobs—but long before he finally moved into the professoriat, his miracle year made it clear that he was a rising star, just coming into the full play of his powers. Thus, in the autumn of 1907 it made perfect sense that he would be asked to write a retrospective, charting the progress of relativity over the previous two years. That kind of invitation is an honor—but this one came with a barb. The request reached him late, leaving him only two months before the December 1 deadline. At first, that didn't seem to be a problem. The larger half of the review went quickly. In four sections Einstein described the application of the relativistic worldview to measurements of time, the study of motion, the behavior of electromagnetic fields, and the implications of Einstein's discovery of the equivalence of energy and matter. That was, in fact, more or less all that he'd been asked to do. His editor had sought a survey of current developments in relativity theory, and he'd written one.

He hung up on one last question, though. The "special" in the special theory of relativity refers to the fact that it is a limited concept. It described the behavior of space and time perfectly for almost all physical situations, with one major caveat: as Einstein understood it at the time, the special theory applied only to motion at a constant rate. That left those circumstances in which speeds change, accelerating or slowing down.* For Einstein, this left an intolerable gap. Acceleration is ubiquitous throughout the cosmos. Most important: anything subject to gravity accelerates.

With all the weight of the Einstein legend between us and that very young researcher—he was still just twenty-eight—stealing time for science in the ebb and flow of patent applications, it's hard to recapture now just how much raw intellectual confidence was required to take this step. Thinking beyond special relativity would inevitably force him to confront the most famous idea— universal gravitation—produced by the most famous physicist in history: Newton. But if his theory really did form a part of the logic of nature, then no idea, not even the most iconic, should be able to escape it.

November 1907. Einstein reported each working day to the Patent Office. He wrote; he thought; he stared into space. Through his window he looked over at the rooftops of Bern. One day—we don't know exactly when—he saw that anonymous roofer; he imagined the accident; and the missing piece, his happy thought,

* Special relativity does apply to accelerated systems; the famous twin "paradox," in which a twin accelerating away from and then returning to the earth ages more slowly than her stay-at-home brother, is an example of a special relativistic analysis of non-uniform motion. But in his initial framing of special relativity, Einstein considered uniform motion, and as he first began to think about generalizing the theory, he continued to think about this distinction between different states of motion.

burst into his mind. The realization that a falling man won't feel his own weight provided the crucial hint that led Einstein to think about gravity along similar lines to those he used to analyze the relativity of time and space. Einstein formalized this insight as the "equivalence principle"—an axiom that would become as important to his thinking as the relativity principle had proved to be in 1905. In its simplest form, equivalence simply holds that a person in free fall—like the imaginary roofer—cannot distinguish between two possible descriptions of his circumstances. He can't say whether he is *falling* under the influence of gravity, or just *floating* in a gravity-free region of space.

In other words, never mind that from where Einstein sits, the roofer is accelerating, speeding up as he plummets. The roofer himself feels no change (until he hits the ground)—only weightlessness, no push or tug of any sort—which is the signature of uniform, inertial motion, the sort the special theory of relativity describes. The two states—free-fall and un-accelerated motion—thus had to be seen as equivalent, both accurate descriptions of the same phenomenon. The inverse is also true: it's impossible for someone (in a closed room) to decide whether the tug she feels as she stands on the floor is that of the earth's gravitational field, or something—say a rocket motor—accelerating beneath her, pushing upward on the soles of her shoes.

Epiphany! This new principle pointed Einstein directly to the essential connection special relativity on its own couldn't make: the link between inertia, which is the measure of mass, and weight, which is mass multiplied by whatever force affects an object. Someone visiting the moon retains the same mass she has anywhere else. But because the moon exerts about 16 percent of the earth's gravitational tug, her weight will be just one sixth what it would measure back home. More generally, weight can be understood as the perception of a change in the motion of any object, no matter whether that shift comes from acceleration or gravity. Free fall produces the same experience as the weightlessness of empty space, far from any source of gravity; acceleration produces the identical perception—that of weight—as does standing still in the grip of the earth's gravitational field. That tumbling roofer and the equivalence principle he inspired told Einstein the bare minimum of what any relativistic theory of gravity would have to include: a mathematical account that could express the still undiscovered physics that established the connection between the inertial path of anyone or thing and the pull of gravity that exerts its hold on all of us as we travel across the universe.

November came to its end. Einstein finished his paper, including its final section on the equivalence principle and what it implied for a relativistic theory of gravity. It was barely a sketch, a hint toward something richer to come. All he really knew at this point was that it was possible to think about gravity in relativistic terms, in the language of inertia and acceleration.

There was one more mundane matter that caught his attention shortly after he sent off his review. Nowhere in that paper did he mention any real-world challenge to Newton's gravitation, no anomalies or suspicious phenomena. Instead, then and for the next eight years, the central issue remained one of theoretical consistency, of reconciling the formal disagreement between special relativity and Newton's ideas. Privately, though, Einstein had a perfectly fine grasp of the tactics as well as the strategy of intellectual combat. He knew that he could win if and only if he could clearly demonstrate that his theory modeled reality better than Newton's. On Christmas Eve he wrote to Conrad Habicht, an old friend, not a physicist, that he was working on a new, relativistic law of gravitation. His aim? "To explain the still unexplained secular changes in the perihelion of Mercury."

Vulcan had long since drifted to the far penumbra of possibility. But now, Albert Einstein, constructing a cosmos on a foundation of relativity, was taking dead aim at the undiscovered planet. From the beginning of his investigation of gravity, Einstein grasped the crucial either/or of Vulcan's existence or absence. He wouldn't mention Mercury again for several years, even in his private correspondence. But he didn't forget it either.

9

"HELP ME, OR ELSE I'LL GO CRAZY"

There is an idea—utterly strange at the time—shot through the fabric of special relativity. In the century since it was first revealed, it has woven itself through the warp of popular culture as much as it flows through formal cosmology. When he first encountered it, though, Albert Einstein was unimpressed. "Now that the mathematicians have seized on relativity theory," he declared, "I know longer understand it myself." The offending mathematician? Einstein's former teacher, Hermann Minkowski. The offending idea? In Minkowski's own words:

"Gentleman, the concepts of space and time which I wish to present to you have sprung from an experimental physical soil and therein lies their strength. They are radical. Henceforth space by itself and time by itself are doomed to fall away into mere shadows, and only a kind of union between the two will preserve an independent reality."

We now call that union "space-time." The old notion that space occupies three dimensions—our familiar height, width, and depth—and time ticks on regardless, Minkowski argued, could no longer hold, not if you take seriously the discovery that one's state of motion affects measurements of both. His response: to propose a world that exists in four dimensions, three of space, one of time, all intertwined with each other.

Most important, Minkowski provided the mathematical apparatus with which to explore space-time. He showed how any two points, any two events—me, sitting as I write this; me rising to grab another mug of coffee—could be combined into a single, absolute picture, one that both observers in motion and at rest could accept as the true description of their differing measurements. The detailed geometrical argument is somewhat complex, but Minkowski's work defined the single path that marks out the shortest track between any two points in four-dimensional space-time. That path is called the absolute interval, one measurement that combines the distances traveled in both space and time between two events. It is a map that does not change. (To make manipulating space-time a little easier, physicists have developed a trick to express measurements of both space and time using the same units. The speed of light provides the ultimate yardstick. How long does a meter last in time? Just so long as light takes to cross that distance—3.3 billionths of a second. How far is a second? It is the distance light travels in one tick of the clock, 186,000 miles, or 300 *million* meters.)

Einstein had always affirmed that while the two observers' measurements would differ, there was only one sequence of underlying phenomena, and all observers would find the same laws of physics behaving in the same ways as they conducted their experiments. Minkowski's accomplishment was to make that sequence explicit—plain for anyone to see. For Minkowski, this was revolutionary. For Einstein, not so much. Four-dimensional geometry he said, was "superfluous erudition."

Minkowski never got the chance to school Einstein on his poor mathematical taste. He died in 1908, just forty-four years old, cut down by appendicitis. Einstein could ignore the implications of the space-time perspective with impunity—for a while—and so he did. There was plenty of other physics to consider, and daily life as well.

That life torqued after 1905, unsurprisingly. While he remained at the patent office for several more years, the inevitable move from a bureaucrat's job to an intellectual's got a hesitant start in 1907, when he became a part-time adjunct at the university in Bern. It took until 1909 to land his first full-time academic job, at the University of Zurich. It was a stopgap appointment, an untenured post that left Einstein at the bottom of the totem pole. But in a little over a year he received an offer of a full professorship, with all the status and security that implied. The difficulty was that the invitation came from the German University in Prague in early 1911—in effect the far frontier of German-speaking Europe. In a common academic's story, he made the leap anyway: the post trumped geography.

Neither he nor his wife, Mileva Marić, ever really liked Prague. Einstein complained to a friend shortly after his arrival that the locals showed "a peculiar mixture of class based condescension and servility, without any kind of goodwill to their fellow man." The city itself was one of "ostentatious luxury side by side with creeping misery on the streets. Barrenness of thought without faith." Even so, he found some compensations. A family connection recalled that he loved sitting in cafés along the river, drinking coffee and talking with friends. The elite fraction of the Jewish community there enjoyed a salon society, and it appears that at least once, perhaps more, he found himself in the same room as Franz Kafka.* For Marić, though, Prague had no virtues at all. It was the old story as told by a friend of the family: "She was left at

---

* Sadly, there is no solid evidence that they ever spoke to each other.

home with the children and became more and more discontented every day."

For Einstein, Prague's saving grace was the chance to work without interruption. No patents to read, and all the freedom universities afford those in their senior ranks. After he first imagined a relativistic theory of gravity in 1907, Einstein had shifted his attention to the vexing riddles of the quantum realm. He made very little progress for the next several years, until his move to Prague left him with little beyond a rueful appreciation of the sheer nastiness of the problem. His office overlooked the grounds of an insane asylum. He was thinking of quanta when he described the inmates he watched from his windows as the "madmen who do not study physics."

So he switched mysteries. Settled in his new home, he returned to the equivalence principle and what it could do to help him extend relativity. He knew that the outline of the idea he had given three years before was inadequate. He now found a way forward, one that required him to think deeply about what gravity might do to light.

To capture the flavor of Einstein's new approach, return to a version of one of his thought experiments, that rocket last seen accelerating in empty space. This time, its designers have cut a window in its fuselage, so that if someone shines a flashlight into the ship at rest, that beam of light will travel straight across to the opposite wall.

Now imagine the rocket takes off, accelerating as it goes. The craft moves just a little in the time light takes to travel from one side of the compartment to the other. As the rocket moves, that flash strikes the far wall just a little lower than the point at which

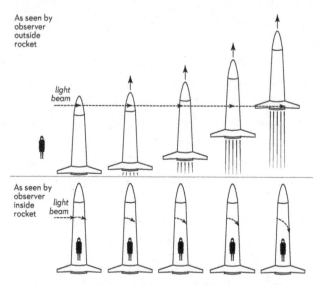

As seen by observer outside rocket

light beam

As seen by observer inside rocket

light beam

In this thought experiment a beam of light shoots through a window in a rocket accelerating upward. To an observer outside the rocket, the light travels in a straight line. But for someone in the rocket, the light enters at one point up the side of a rocket, and hits the other side farther down, following a curving track to get there. Within the accelerating frame of reference, light bends. By the equivalence principle, the same thing happens in a gravitational field.

it entered through the window. To an observer within the rocket, light is bending, actually curving downward. Accelerate more, and each ray of light bends more sharply. That's under acceleration—and if you accept the equivalence principle, then, as acceleration bends light, so must gravity.*

---

* Einstein did not simply settle for pictures. In this, his first paper on gravity in several years, he started to generate some hard numbers. For the bending of light, he

The next stop followed logically. Given the relationship between light and time in special relativity, it came as no surprise to Einstein that the behavior of light in this new setting would lead to an effect on the flow of time as well. To simplify his reasoning, return to the rocket. Imagine there's a clock set in the nose of the ship, at the top, and a second clock at the bottom bulkhead, back by the engines. At rest, the two clocks remain synchronized by a flash of light—a signal—that the clock at the bottom sends to the clock at the top, once per second. Both clocks keep good time, and—more important—the *same* time. It gets interesting, though, when the rocket's engines start and the ship begins to accelerate.

While the clock at the bottom flashes light, the rocket moves, faster and faster. In the time that the flash of light takes to travel from the bottom of the ship to the clock at the top, the ship has risen, just a bit. The distance the signal must travel grows, which means, of course, that the bottom clock's light signal takes more time to reach the top clock than it does when the rocket is at rest.

---

calculated how much a mass the size of the sun would divert a ray of starlight just scraping past its edge. He came up with a figure, .87 of an arcsecond, that was, not quite coincidentally, the same value that Newton's theory generates. The number was wrong, as we shall see, but Einstein believed it for the next four years. By way of scale: a circle can be divided up into 360 degrees. Each degree splits up into sixty minutes (or arcminutes, or minutes of arc, depending on such niceties as whether one speaks English or American), and each minute can be further divided into sixty seconds of arc. .87 seconds of arc is a small but not impossibly tiny number. If one were to draw a line from the Bulfinch dome atop the state capitol in Boston to Times Square, that line would stretch approximately two hundred miles. Coming to rest at the half-price Broadway ticket booth at the heart of the square, a deviation of .87 arcseconds would land one about six feet to either side of the line queuing for the chance of the latest hot seat.

This thought experiment turns on similar reasoning to that which Einstein used to analyze simultaneity. Acceleration increases the distance each successive signal from the bottom clock must travel to reach the top clock. An observer sitting beside that top clock would thus observe the lower clock to be running slowly. Again, by the equivalence principle, the same would hold true under gravity, with time slowing as a gravitational force grows stronger.

The same holds for the next flash, and the one after that. Anyone checking would see that each light pulse arrives in just a little more than a second, as ticked off by that top clock. That means that the clock at the bottom is running slower than the clock at the rocket's nose. Once again, by the equivalence principle, clocks in a gravitational field must behave in exactly the same way that the rocket's clocks do. A clock placed where gravity tugs more strongly, closer to the center of the earth, must run more slowly than one perched a little farther from the earth's center. The tick of time runs more slowly on the flat plain around Berlin than it does on top of the mountains near Zurich that the young Einstein used to climb.*

With that, Einstein fought through to the last step in his chain of reasoning. The invention of special relativity had already altered the conception of time. No longer an absolute, it became simply that which any given clock actually measures, given its motion relative to an observer. The one consolation was that the different clocks could be reconciled, using Minkowski's mathematics of space-time, to bind one observer's findings to another's. But with his work in Prague, Einstein complicated those limited certainties. If gravity affects clocks, that means that time must vary from place to place. It bows to circumstance, whether one finds oneself at the Dead Sea or on Everest, or even merely in the basement or on the third floor. That each place, every place, has its own unique flow of time was a new vision, and not a comfortable one—not then, and not now. But by no later than the middle of 1911, Einstein saw where any extension of relativity theory must lead. Gravity bends time.

---

* Richard Feynman, in the final lecture reprinted in *Six Not-So-Easy Pieces*, presents one of the best and most straightforward discussions of the ideas discussed here, but the clock-and-rocket-ship example has been around for quite some time. Feynman does give a nice sense of perspective, providing, for example, a number for the amount that time slows as one climbs to weaker and weaker heights in the earth's gravitational field. Every twenty meters up alters the frequency of light, and hence the measurement of time, by two parts in a thousand trillion ($2/10^{15}$).

. . .

With that insight, Einstein came to realize that there was a feedback loop in his emerging view of gravity. Getting there took a complex, subtle chain of reasoning. In his theory, as in Newton's, gravity performs work—in the physicist's sense—making objects move. In Newton's version, all the force required to do that work depends on the amount of mass involved; that's what his famous equation means. But Einstein knew from $E = mc^2$ that energy and mass are equivalent, two faces of that single entity, mass-energy. The next thought was in some ways obvious after the fact—but at the time it represented a breakthrough. Any change in the amount of potential energy contained within a gravitational system would alter the total amount of mass-energy present—and hence the intensity of the gravity acting on the objects involved. That is: Einstein now realized that gravity can impose its own effect on itself, that every change in the conformation of the system alters the system's gravitational behavior. That, finally, forced him to confront the fact that his mathematical task had just grown much more difficult: any theory of gravity consistent with special relativity would have to model that interplay between energy and mass—which, in technical terms, means that it would have to account for a nonlinear process.

That was a blow, at least for Einstein's hopes of coming up with a mathematically simple theory of gravity: nonlinear equations are notoriously more difficult to solve, so much so that it's a standard tactic to try to convert nonlinear problems into linear expressions—which Einstein now knew he could not do. Still, this was the observation that allowed him to enter the next phase, to go beyond what he'd learned about the behavior of gravity in the wild to the fundamentals, a law that would allow him to model the way a relativistic gravitation had to work. Throughout his first winter and into spring in Prague, he made little headway. He told one friend in the spring of 1912 that he had come up against enormous obstacles, and another that "I have been working furiously on the gravitation problem...every step is fiendishly difficult."

Here, at last, Minkowski's description of four-dimensional space-time came to the rescue. Minkowski's chief motive in creating that concept lay with his desire to clarify the implications of special relativity. His scheme retained one vital characteristic held over from earlier ideas: its four-dimensional cosmos that served as the container for whatever mass and energy might be doing within it. It was the stage on which history happened, unmoved and unmoving in itself.

Einstein's great advance came when he thought about what it had to mean that acceleration and gravity affect the flow of time: *space-time* would have to bend too, as one of its dimensions (time) flexes under the influence of gravity. With that realization, Einstein's thinking took on the elegant sweep of his best work. His new catechism: gravity is a property of matter and energy together (not matter alone, as in Newton's view)—and gravity bends time. Taken together, those two facts led to this conclusion: the total amount of mass and energy determines the strength of the gravitational field in any particular location, and hence the amount any given region of space-time will flex. That warping of space-time in its turn has to affect the paths that matter and energy may take through the cosmos. Space-time is no stage, merely the box the universe comes in; rather, Einstein now realized, it is active, dynamic, shaped by what it contains. As he later put it, he had at last grasped a crucial truth: "the foundations of geometry have physical significance."

That vision didn't mean Einstein was done. But this was *the* essential step. With it, he'd managed to take a pure physical insight—gravity and acceleration are equivalent—and refine it to the point where he now knew what a full, rigorous mathematical account would have to include. He still didn't know enough to get there himself; he didn't even know what sub-discipline within the math universe would meet his needs. But he knew a guy.

Marcel Grossman was both a first-class mathematician and one of Einstein's oldest friends. They had met as undergraduates at the ETH in Zurich, and Grossman had been known to lend his classmate his notes when Einstein played hooky. Their reunion came when Prague, a backwater, found it couldn't hold on to a talent like Einstein. The ETH sent out feelers early in 1912, and by summer, the deal was done. Less than two years after leaving Switzerland, Einstein returned in triumph as the professor of theoretical physics at one of the top technical universities in Europe. Grossman was already there as professor of mathematics. At their reunion Einstein begged: "Grossman, you must help me, or else I'll go crazy."

Grossman did. He already knew what Einstein needed: a way to escape what had for two thousand years seemed to be the only way to map the shape of nature: Euclidean geometry. It's almost impossible to overstate the hold on natural philosophy Euclid's book, *Elements,* possessed for most of its history. In more than two millennia no one has found an error in its analyses of planes, surfaces, and solids. The shortest distance between two points on a plane is a straight line, through a point not on a line there is no more than one line parallel to the line, the angles of a triangle add up to 180 degrees. All this and more seemed to be necessarily true—not only in the *Elements*, but out here, in the real world.

And then it wasn't. In the early and mid-nineteenth century some of the most audacious thinkers in mathematics discovered they could modify one or another of Euclid's assumptions, called axioms or postulates—those statements taken to be so obviously true that they required no proof. They found alternate geometries, just as consistent as Euclid's, but ones in which, for example, no parallel lines exist. Grossman told Einstein that the version invented by Bernhard Riemann could serve his needs: it could analyze how to make measurements at any point on a smoothly changing curved object. When Riemann had created his system, he was thinking like a mathematician, one who focuses on ideas, not things. But for Einstein, this was a revelation: This strange, unsettling vision allowed him to play with space, treating it as something in which matter and energy would move, not on straight lines, but on curving paths.

Most important, he could now answer the crucial question: what, exactly, *is* gravity? Clearly not Newton's instantly felt occult force-at-a-distance. Instead, in his still-forming analysis, gravitation is built into the geometry of space-time. Formally, gravity is the local curvature of space-time, the particular shape given to it by concentrations of mass-energy, like the earth or the sun. Mathematical analyses of such dents reveal the precise relationship between the distribution of matter and energy and the nearby shape of space-time. Objects navigating the cosmos—planets in orbit around a star, moons around a planet—are not mysteriously dragged along those paths. Rather, they simply follow the shortest route available to them around the dings and dips in space-time produced by all the matter and energy in their vicinity.

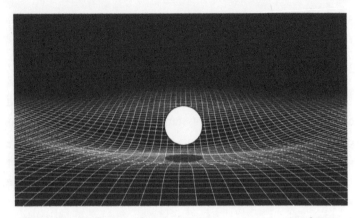

The classic visualization of gravity-as-geometry sees a massive object—like a star—stretching and deforming space-time imagined as a kind of rubber sheet. It's an imperfect analogy, but it gets to the idea.

A problem remains, at least as far as gaining an intuitive sense of how the shape of space-time generates what we all experience as a force, the one that splashes wine on the floor when you tip over your glass. To get a sense of what's going on, imagine a vast, seemingly featureless plain. It's so flat that anyone living there can only perceive two dimensions, length and width, with no discernible changes in altitude. Go for a walk—say, on the most direct track between your home and a distant village—and you find that after a mile or so your steps come harder. It takes a little more effort to keep going. You begin to puff and labor. You clearly sense that you're being tugged by *something*—a force you could call gravity. It pulls at you as you walk along what you are sure is a straight line. To anyone watching from a vantage where they can perceive three dimensions, not two, there is a simpler explanation. What feels like a mysterious force is simply the result of taking the shortest path up a hill.

That is: the "gravity" felt by a hiker on that empty plain is nothing more than the measure of a curvature of space, a rise the walker cannot actually see. The analogy is not perfect, as it only deals in space, not time. But it gets to the nub of the matter: we inhabit a locally curved region of space and time created by the mass of the earth. The weight we feel as we stand by our beds in the morning is the sensation of our daily slide down a well in space-time, a warp bending down toward the center of the earth. That sensation is born of the geometry of experience, an exercise in space-time dynamics that, every day, holds our feet to the floor.

It was all there by mid-1913. Einstein had a physical picture, and he was finally working on it with the right mathematical tools. Most important, he had a model, both in the formal quantitative sense and as a mental picture, a way to imagine this still-bizarre notion of gravity as geometry-in-action. Einstein was confident—more, almost celebratory. He wrote to his fellow physicist Ludwig Hopf at the beginning of his work with Grossman, "It is all going marvelously with gravitation. If it isn't all a trick, I have found the most general equations."

It wasn't *all* a trick. It just wasn't quite right. It had taken Grossman and Einstein almost a year to produce what they called a "Draft of a Generalized Theory of Relativity and of a Theory of Gravitation." That title was fair. For all the time they'd spent on it, that paper was genuinely a draft. As published it contained several important errors, some simply slips of calculation, but others flowing from the fact that Einstein hadn't yet truly mastered how to marry his physics to the dense and difficult mathematics Grossman had taught him.

But the idea itself was almost there, as Einstein knew, or at least felt very deeply. It was close enough for him to see that it could be tested. The theory made one clear prediction: light as well as matter would have to follow the contours of space-time, which meant that a ray of light passing close to the edge of the solar disk would bend round that gravity well created by the sun's mask. The effect was big enough to be detectable, Einstein realized, but only during a total eclipse of the sun. Under the new theory that deflection would be .87 seconds of arc—a number well within the reach of experienced eclipse observers.

Einstein kept to himself the other possible validation. But in a document that surfaced more than three decades after his death, it turns out that he and Michele Besso, an amateur of science who was his dearest friend, tried to model a single specific circumstance: the problem of Mercury—what would happen to its orbit, rolling around the steep slopes deep down the well in space-time created by the sun.

Their calculation, mostly in Einstein's handwriting with corrections and a few more substantial contributions from Besso, offers some guilty pleasures. Einstein made a couple of elementary blunders—multiplying the mass of the sun by an extra factor of ten, for example—the kind of careless mistakes that come as a comfort to us mere mortals. He also made at least one more serious error in persuading himself that an approximate solution for another, more abstract test of the theory was valid when it was not.

At the same time, the work offers a rare window into the act of scientific thinking, a very different kind of picture than the ones to be found in the static, artificial picture of discovery in most published results. Einstein is groping here, both trying to master the big ideas in an unfamiliar and difficult body of mathematics and to develop the specific techniques to use that math to generate detailed accounts of the behavior of matter in motion. The calculation for the orbit of Mercury contains a real advance. In it, Einstein developed a valid method to analyze the motion of a planet moving through curved space-time. But given the still-hidden flaws in the Einstein-Grossman version of gravitation, that techni-

In this letter to the American astronomer George Ellery Hale, Einstein seeks advice about measuring the deflection of starlight around the sun.

cal achievement didn't help just yet. Instead, when they worked through all the steps, Einstein and Besso found that they still could only account for just eighteen of the forty-three arcseconds per century they needed.

On the face of it, that was as much a failed result as any of the Vulcan sightings and misidentified sunspots were under Newtonian gravitation. Einstein himself seemed to respond to the problem just as Vulcan's partisans had decades before: the failure to confirm his view mattered less than the theory itself. If a relativistic account of gravity made sense, if it retained its logical and explanatory power, a single missing confirmation was hardly reason to abandon it.

Still, such a result was certainly nothing to advertise. Einstein never published this particular exercise. Instead, he just kept going. He knew that Newtonian gravity was inadequate as a matter of principle—the conflict with special relativity that could not be wished away. He could feel the logic of a relativistic theory of gravity accumulating with each new insight. If it was not yet fully formed, he was still convinced that it represented the only reasonable path forward—so much so that he was prepared to subject it to the most public of tests. The next total eclipse of the sun was scheduled for August 21, 1914. Russia's Crimean peninsula, jutting out into the Black Sea, offered prime viewing conditions. There, astronomers would have their first chance to check the theory's major prediction: that a ray of light from a star grazing the edge of the sun would be deflected by .87 arcseconds off its usual path as it careened around that patch of sharply bent spacetime. The symmetry is obvious: Vulcan, of course, had been sought and seen and unseen again in such conditions.

There was, however, an obstacle in the way of testing Einstein's account of gravity (and hence of Mercury's motion) at the 1914 eclipse—one that had nothing to do with his scientific argument. Einstein himself had risen far since his miracle year, but as a professor in Zurich he had no real hope of finding the funding to mount an expedition to Russia. In July 1913, that ceased to be a problem. Two men from Berlin came to call on him, Max Planck and Walther Nernst, both future Nobel laureates. They made Einstein an unprecedented offer: if he would abandon the Swiss and join them in Germany's imperial capital, he'd gain a truly impressive salary, a faculty appointment with no teaching requirements, and membership in the Prussian Academy of Sciences.

Despite such temptations, there were plenty of reasons for Einstein to reject the invitation. Einstein loved Zurich, and had renounced his German citizenship at the earliest opportunity more than a decade earlier. But this offer said, in effect, that if he came to Berlin, he'd arrive as first among equals within the single most impressive collection of scientific talent in the world.

Einstein took a day to think it over, but it really was too good a deal to refuse. It also came with an implied bonus: his Berlin hosts wanted him to be happy. In practice, that meant he could now draw on enough money to fund an eclipse-spotting team.

Einstein left Zurich in March 1914. He crisscrossed Europe for a few weeks visiting physicist friends before settling into the Berlin suburb of Dahlem in April. His marriage didn't survive the move. Marić and their sons returned to Zurich in early July, and Einstein, though he wept as they left, swiftly returned to what he clearly loved best, thinking about physics unburdened by (as he called it) "the merely personal."

The eclipse expedition was duly organized. The Prussian Academy paid part of the total—the expected sweetener to charm their newest recruit—and the patriarch of the Krupp family covered the balance. Erwin Freundlich, a young astronomer and Einstein enthusiast, selected four camera-equipped telescopes and recruited two companions. They left Berlin for the Crimea on July 19. No one seemed to think that an event three weeks earlier would impinge on such a purely disinterested quest. What possible concern could it be for stargazers from Germany, bound for Russia, that on the 28th of June, 1914, in the Serbian city of Sarajevo, an Austrian archduke had got himself shot?

10

"BESIDE HIMSELF WITH JOY"

July 1914.

In Berlin, a delight.

Looking back years later, long after Europe's Great Powers consummated their four-year murder-suicide pact, it seemed life had never been sweeter. Journalist Theodor Wolff crystallized those last prewar weeks in the craze of the day: "The Berlin public discovered a new passion," he wrote. "After the one-step and the two-step had done their duty, the new miracle was called Tango."

The pleasures of the season persisted for a good while after a Serbian radical killed the Austrian archduke Franz Ferdinand and his wife, Sophie, on the 28th of June. The next morning, the first printing of the *Frankfurter Zeitung*—Germany's most serious newspaper—covered the murder. The second edition, though, returned to conventional summer fare, including an article urging public support for the Berlin Olympic games slated for 1916. Three weeks later, the press still covered what seemed—and was!—a perfectly normal summer. On July 21 the *Berliner Volksblatt* told its readers how to get a tan, while warning of the possibly immoral costumes that produced the best results. Perhaps the most striking proof of the incomprehensibility of war in those last weeks appeared in the classified ads: during that summer the *Frankfurter Zeitung* carried for-sale notices aimed at their upscale audience—for vacation property in Russia.

It was thus utterly ordinary that a German scientific expedition should show up on Russia's Black Sea coast in the last week of July, just ahead of the eclipse that would pass that way on August 21. Erwin Freundlich and his colleagues had come to measure what Albert Einstein had told them to look for: the bending of starlight around the sun. They met up with a group of astronomers from Argentina who were, in a coincidence no fictional account would tolerate, just about the last holdouts, a group seeking to photograph a hypothesized planet inside the orbit of Mercury planet—that old friend, Vulcan.

Then, in less than a week, sunlit Europe, dancing Europe vanished. On the 30th of July, Nicholas, Czar of all the Russias, committed his empire to full military mobilization. Germany called on Russia's ally France to remain neutral. France refused, mobilizing on the 31st. On August 1, the German ambassador to Russia delivered a document to the foreign ministry in St. Petersburg: the formal declaration of war.

That night, a small German force entered neutral Belgium. Germany's patrols crossed into France on the 2nd and delivered the now redundant paperwork the next day. Finally, at 11 P.M. on August 4, His Majesty's government made its decision, informing the Kaiser's ambassador to the Court of St. James's that a state of war existed between the British Empire and the German Reich. Among the least noted consequences of that rush to war: three German scientists had just become enemies to their hosts. Freundlich and his companions were arrested and interned, their equipment seized.

In the end, it didn't matter. The eclipse was a tease. Clouds gathered just before totality and cleared beautifully only after it passed; no light-bending observations could have been made. The German astronomers were lucky. Detained only briefly, they were swapped for Russian officers in one of the first prisoner exchanges of the war. To Einstein's relief, he was able to welcome them back to Berlin by the end of September.

That was one of Einstein's few consolations in that miserable time. He never reconciled himself to the shock of the war—not merely the fact of battle, but the naked joy that everyone, it seemed, took in the fight. "That a man can take pleasure in marching in fours to the strains of a band is enough to make me despise him," he wrote years later. "Heroism on command, senseless violence and all the loathsome nonsense that goes by the name of patriotism—how passionately I hate them." What was worse, in Einstein's eyes, was that the extraordinary collection of scientific minds that had lured him to Berlin turned out to be as war-drunk as any mob in the street.

The most potent symbol of this almost-personal betrayal was Einstein's closest Berlin friend, Fritz Haber, who would later win the Nobel Prize for chemistry. With the start of the war, he shifted his lab to a near-total military focus. The element chlorine caught his attention, as he pursued what he hoped would be the weapon to end the war to end all wars: lethal poison gas, illegal under prewar agreements, which he now proposed to deliver to the German General Staff.

Haber managed to produce battlefield-ready chemical munitions early in 1915. That spring, cylinders loaded with chlorine were shipped to the Western Front, to be deployed on the Ypres battlefield in Belgium. After weeks spent waiting for the winds to blow steadily from east to west, April 22 offered made-to-order conditions. As dusk approached, German soldiers released 168 tons of chlorine gas along a four-mile front.

In its path stood three divisions: one Algerian, one territorial (the equivalent of a national guard unit), both under French colors, and one Canadian. A green-tinged cloud advanced, drifting, rolling, stretching across the muddy wreck of no-man's-land. The Germans waited until the billowing mass of chlorine reached the Allied lines. The effect was everything Haber had hoped for: hundreds of soldiers "were thrown into a comatose or dying position," as General Sir John French reported. The Algerians broke, leaving a gap in the line. The Germans advanced, taking two thousand prisoners and a number of artillery pieces—until the Canadians reformed and reasserted the relentless stasis of trench warfare.

Such a "victory" was a fiasco typical of the brutal stasis that had already gripped the Western Front. Achieving complete surprise, using a weapon for which their enemy had no defense, the Germans gained exactly nothing. The Germans had no reserves available to exploit an advantage they possessed for the briefest of moments. The western Allies soon retaliated with gas munitions of their own, with no greater tactical success than the Germans.

Gas was and remains a terror weapon, ineffective against prepared and similarly armed opponents. Both sides would continue to launch poison attacks throughout that "great" war—and Haber himself persisted for years, seeking the one magic formula that would at last deliver a strategic breakthrough rather than just a particularly nasty increment to the misery of trench warfare. He never found it.

To Einstein, this was simply madness. "Our whole, highly praised technological progress," he wrote in what is now one of his most famous aphorisms, "and civilization in general can be likened to an axe in the hand of a pathological criminal." World War I broke something in Einstein, destroying forever the faith he'd affirmed until the fall of 1914—that there was a genuinely supranational, disinterested elite of the mind, united by what he most valued, the study of "this huge world, which exists independently of us human beings and which stands before us like a great external riddle." As a youth he "soon noticed that many a man whom I had learned to esteem and admire had found inner freedom and security in devoted occupation with it." Those were the people he thought he'd joined in Berlin, and now, mere months after his arrival, they had (as he saw it) abandoned him.

Yet he stayed. He didn't have to. He still held Swiss citizenship, and he was able to cross the border between Germany and Switzerland during the war. Zurich had long been his favorite city, and as the British naval blockade came to bite, wartime Berlin would become not just politically grim but flat-out hungry. None of that seemed to matter. Berlin, it turned out, had one unsurpassed advantage. If all his colleagues were consumed by the passion and labor of war, at least they left him undisturbed. His wife and children were gone. He lived alone. He kept his office in Haber's chemistry institute, while ignoring that which he disdained going on all around him. Uninterrupted, he could think.

So, once the initial shock of August wore off, Einstein got back to work. On October 19, he delivered the first of two lectures to the Prussian Academy. He spoke not of war, but of gravity, introducing what he now saw as a substantially complete generalization of relativity.

In those talks, Einstein argued that his new theory presented not just the solutions to certain problems, but a whole new way of thinking. He explained the significance of his prior work on non-Euclidean geometries, those mathematical systems in which parallel lines could meet and space warps. Such concepts were not just abstract toys for clever mathematicians but, rather, he said, should be understood as actual candidate descriptions of the real world. Using them, it became possible to contrast competing ideas: Newton's account of gravity as a force and Einstein's own, in which the shape of space-time determines how objects move.

It was an utterly foreign message for that audience. To be fair, Einstein admitted that his theory was not yet complete, even if it seemed to him logically sufficient. But the claim that geometry was destiny was simply too bizarre a thought for most. Even allowing for such difficulties, though, it's remarkable that none of his German colleagues, despite all the effort they'd expended to lure Einstein into their midst, paid any real notice to what they'd just heard: first, that Isaac Newton was wrong about gravitation, and second, that to get gravity right, physicists had to reimagine fundamental assumptions about the behavior of the universe. Einstein told them to their faces, twice, and he published the argument as a fifty-five-page article in the proceedings of the Prussian Academy. When his article appeared, Einstein did receive a few letters from foreign researchers exploring the corners of his ideas, but nothing fundamental. No one in Berlin went even that deep.

Einstein was unsurprised. The year before, the dean of German physics, Max Planck, had warned him not to tackle gravity. It was too hard a problem, he said, and "even if you succeed no one will believe you." Science may celebrate the triumph of the better idea. Scientists don't, not always, not immediately, not when the strangeness involved takes extraordinary effort to embrace.

Einstein ignored both Planck's advice and the apathy of his colleagues. His October lectures contained the most complete account of his ideas about gravitation he could manage. A problem remained with the interpretation of his equations: in certain circumstances his 1913–14 theory violated a key claim of special relativity, that every observer in motion relative to each other must come up with the same description of an event in the mathematical language of space-time.

For the time being, Einstein didn't think that reversal invalidated the generalization of relativity to motion under the influence of gravity. *That*, he was convinced, had to be right. Nonetheless, the loss of invariance was certainly worrisome. For the moment, though, he had reached his limit. If there were errors, he could not—yet—recognize them.

The first year of the war came to an end, marked by the beautiful and melancholy Christmas Truce spontaneously declared by ordinary soldiers on both sides of the Western Front. Einstein gave the first months of 1915 over to various distractions, including an utterly unsuccessful jaunt into the wilds of experimental physics. He thought about the battlefield, and his hatred of it, and later that year distilled his anger into his first public statement on war and peace.

Then, after about eight months of letting his mind wander, he persuaded himself to confront gravity one last time.

David Hilbert was the most influential German mathematician of his day. He remains famous both for his own work and for his list of twenty-three "Hilbert Problems"—questions unsolved as of 1900 collected in a sally aiming to shape twentieth-century mathematical research. For Einstein, Hilbert, a professor at the University of Göttingen, had special significance: he was one of the very few first-rank mathematicians who took an interest in his work. He did what no one in Berlin had thought to ask: he invited Einstein to give a series of six in-depth lectures on the current state of his thinking.

In those talks, delivered in late June and early July, Einstein still believed that his work of the previous two years was largely satisfactory. Never mind that he couldn't quite bring his new general theory into perfect agreement with the special one, nor that the equations he announced in 1914 did not yield the correct orbit for Mercury. The transformation of gravity into geometry was, he remained convinced, correct in all but details. That was what he said in Göttingen, and those lectures had, he felt, left Hilbert ready to accept the necessity of his new approach to gravitation.

He was right. Hilbert believed him, so much so that he began working on his own version of a theory of gravity compatible with special relativity. It's not clear when Einstein realized he had a competitor working on the same problem he'd wrestled with for eight years, nor when he decided to reexamine his work with a newly critical eye. It can't have been later than September 30, when he told his friend and supporter Freundlich that his theory was in deep trouble. The trigger was his sudden grasp of a question he'd faced as far back as 1912, one posed within an idealized representation of a rotating system. Analyzing acceleration in that rotating frame of reference spat out results that seemed to violate the equivalence of acceleration and gravity—the founding principle of the entire effort. It was, he told Freundlich "a blatant contradiction"—fatal for the theory as it stood. That same flaw, he added, would explain the theory's inability to generate an accurate orbit for Mercury. Worst of all, he couldn't see a way forward. "I do not believe that I myself am in the position to find the error," he wrote, "because my mind follows the same rut too much in this matter."*

What a cry for help! Freundlich didn't reply (or at least no letter survives), and in any event he was no deep theorist. It didn't matter. Sometime over the next week, Einstein figured out how to proceed. As soon as the solution started to form in his brain, he went almost completely silent. From the 8th of October, he wrote just four letters—two brief notes to organizations, one to a friend in Zurich, mostly about family, and one substantial scientific memo to an older researcher whom Einstein greatly admired, the Dutch physicist Hendrik Lorentz, in which he discussed some of his emerging ideas about gravitation. For the rest, Einstein seems to have devoted all his time to thought and calculation.

It was, he would recall, the most intense labor of his life.

Even though there is no record of the detailed sequence of his work over the next six weeks, the broad outline of this final push is clear. In one of Einstein's notebooks from his prewar collaboration with Grossman, he had worked through some ideas that added up to an almost-complete version of what would become his final result. In 1913 he'd played with them briefly, and then

* I hope it doesn't make me a terrible person to say how much I love this admission. Einstein himself never had any doubt of his own fallibility, but it does help the rest of us to be reminded that even the best minds can chase their own tails. At least it helps me.

rejected them. Now, two years on, he was ready to take another look.

Albert Einstein in Berlin in 1914

For the next several weeks he used that earlier approach to solve the biggest objection to his prior work, to show that acceleration and gravitation remain equivalent under all circumstances. October ended. Einstein was almost all the way there, and he knew it. On Thursday, November 4, he made his way along Unter den Linden and presented the Prussian Academy with the first of four updates on his progress. He still hadn't fully fleshed out the new theory—the biggest piece missing was the final, correct equation for the gravitational field. He hadn't calculated a specific result for any of the key tests of the idea. But the work made sense now; it was, at last, internally consistent. Equally important, he showed that an approximate solution to his field equations reproduced Newton's laws of motion—just as it must, given how spectacularly Newton's system of the world had accounted for almost all of the motions of the solar system.

The next Thursday he went back to the Academy with an update that contained what turned out to be a mistake, one he would fix two weeks later. He returned home and continued to think and calculate. A week passed. The theory was becoming robust enough to compare to reality. Never mind the problems remaining in his math, he would write. They were formal concerns, not physical ones, and, he wrote, "I am satisfied" not to worry about the fine print "for the time being."

Instead, he turned to the heart of the matter: he placed "a point mass, the sun, at the origin of the co-ordinate system." Next, he calculated the gravitational field such a point mass would produce. Analyzing events within that field, the first novelty appeared almost immediately: the bending of starlight around the sun, just as in his earlier theory—but with this difference: a ray of light passing through the sun's gravitational field would deflect by 1.7 arcseconds, double the number his 1913 theory predicted.

That was the prelude, an undercard bout. The main event was at hand, a demonstration that his theory captured something of reality that no other idea could explain. Two weeks earlier, he'd shown that Newton's gravity emerged naturally from a first approximation—a kind of low-resolution image—of his new mathematical account of gravitation. He repeated the analysis, and extended it to the obvious next question: what emerges from solutions to a second order, effectively, an exploration at higher resolution? A page of mathematical argument followed, yielding a new equation with one term altered from the Newtonian approximation.

Seven more steps, and he had it: an equation that he could use to analyze the orbit of a planet tracking round its star, still sitting in the center of his coordinate system. If he knew by observation just a handful of parameters, he would then be able to generate a prediction for the perihelion advance for any body in orbit around its central star.

November 1915, between the 11th and the 18th.

Einstein gathers the data for Mercury. He writes down its period. He enters its orbital parameters and its nearest approach to the sun. He injects the speed of light into his mathematical apparatus. He does the arithmetic. As he completes each step of the operation, numbers emerge. He peers at the result....

November 18, 1915.

Masking his emotions behind the required decorousness of scientific communication, Einstein revealed almost no sign of any excitement in his presentation to the Prussian Academy. "The calculation for the planet Mercury yields," he told his audience, "a perihelion advance of 43 arc seconds per century, while the astronomers assign 45" +/- 5" per century as the unexplained difference between observations and the Newtonian theory." Belaboring the obvious, he added that "this theory therefore agrees completely with the observations."

Such neutral tones could not conceal the explosion thus detonated. Decades of attempts to save the Newtonian worldview were at an end. Vulcan was gone, dead, utterly unnecessary. No chunk of matter was required to explain Mercury's track, no undiscovered planet, no asteroid belt, no dust, no bulging solar belly, nothing at all—except this new, radical conception of gravity. The sun with its great mass creates its dent in space-time. Mercury, so firmly embraced by our star's gravitational field, lies deep within that solar gravity well.* Like all objects navigating space-time, Mercury's motion follows that warping, four-dimensional curve . . . until, as Einstein finally captured in all the abstract majesty of his mathematics, the orbit of the innermost planet precesses away from the Newtonian ideal.

It was said of Newton that he was a fortunate man, because there was only one universe to discover, and he had done it. It had been said of Le Verrier that he discovered a planet at the tip of his pen. On the 18th of November, 1915, Einstein's pen destroyed Vulcan—and reimagined the cosmos.

In private, among friends, Einstein allowed himself to feel his victory. The equations themselves had simply cranked out the correct orbit. Put the numbers in, and out pops Mercury—as if, to use his own word, by magic. Einstein felt all the pure wonder of that perfect match between theory and reality. Working at his desk, some time in the week before he rose before the Academy, the correct answer appeared as he cranked through the final

---

* Farther out, the sun's influence on local space-time moderates, and the orbits of the other planets approach the Newtonian approximation (though with modern instruments, it is possible to detect a relativistic component in the orbits of several solar-system bodies).

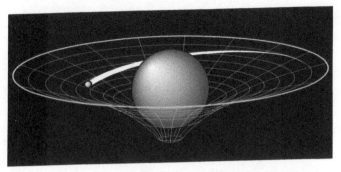

The sun's great dent in space-time dictates the path Mercury must take—an ellipse, precessing around its major focus in precise agreement with observations first quantified by Le Verrier in 1859.

steps. That was when, he told a friend, his heart actually shuddered in his chest—genuine palpitations. He wrote that it was as if something had snapped within him, and told another friend that he was "beside himself with joy."

Much later, Einstein tried again to describe what he felt at that first, private instant of great discovery. He couldn't. "The years of searching in the dark for a truth that one feels but cannot express, the intense desire and the alternations of confidence and misgiving until one breaks through to clarity and understanding," he wrote, "are known only to him who has experienced them."

Postscript

"THE LONGING TO BEHOLD . . .

PREEXISTING HARMONY"

Three weeks into the era of general relativity, Vulcan was gone forever. After half a century in which it had been at once necessary and absent, it was finally revealed to be pure fiction. Its repeated "discovery" was nothing more than an object lesson in how easy it is to see what ought to be rather than what is.

Still, Einstein's dispatch of a ghost planet bedeviling Mercury didn't quite finish the job he'd set himself. On November 25, 1915, he returned to the center of Berlin for the fourth Thursday in a row. At the Academy, he rose to present his final theory of gravity. No errors remained, no unnecessary assumptions, no special observers. The work was done. He came to a close, paused, perhaps, to make conversation as necessary, and left.

The glow lingered. A few days later, he told Besso that he was "content, but a little worn out." He allowed a bit more to show through in a letter to a physicist friend. Study the equations well, he wrote, for "they are the most valuable discovery of my life." In its most compact form, that discovery boils down to a single equation, now called Einstein's, a single line of symbols from which all else flows:

$$G_{\mu\nu} = 8\pi G T_{\mu\nu}$$

On one side lies space-time; on the other lies matter-energy: together, the two halves of the universe. The equation defines their relationship—most simply, it shows how matter and energy together tell space-time (the universe) what shape to be, and how space-time tells matter-energy (all that the universe contains) how to move. The result is a universal theory, an account of the shape of the cosmos, its evolution, and even, potentially, its ultimate fate.

In late 1915, almost all the world had no idea that an intellectual revolution had just been won. World War I had three more bitter, brutal years to run. Even those few who truly grasped the implications of general relativity could not evade the war's reach. Karl Schwarzschild devoured general relativity as soon as Einstein's published lectures reached him, at the Eastern Front. In February 1916, still on active service, he worked out the first exact solution to Einstein's field equations—a result that pointed to what we now call a black hole. Einstein was unconvinced that such a weird possibility would have any real physical significance, but he presented the paper to the Academy as Schwarzschild's proxy.

That was Schwarzschild's last meaningful scientific accomplishment. In the filth of the battlefield, illness was almost as great a threat as a bullet—and that spring he contracted a rare skin disease. It took him two months to die. Einstein privately deplored Schwarzschild's too-patriotic politics—but publicly eulogized the loss of his ferociously powerful mind.

Walther Nernst was another Einstein colleague enmeshed in the war. The chemist had made the pilgrimage to lure Einstein away from Zurich in 1913. In August 1914, he launched into a more farcical journey, having his wife drill him in proper military bearing before racing westward in his private motorcar to see if he could serve as a courier for German troops on the road to

Paris. A fifty-year-old bespectacled professor being of no great use on the front, he soon returned to Berlin. But he sent his two sons into the army, and by 1917, both were dead. The revulsion Einstein felt for war fever could evoke brutal contempt for his thus-afflicted friends, but some disasters were too much, even for him, to pass over with a detached "I told you so." On learning of Nernst's boys, he said, "I have forgotten how to hate."

Such deaths stand in for millions. In Europe's charnel house there was almost no spare intelligence left in the scientific world to think about the geometry of space-time. That placed general relativity in an awkward position: solving the problem of Mercury's orbit was a powerful argument that the new theory was correct. But the ultimate test of any new result lies with its predictions: whether it reveals some previously undetected phenomenon to be confirmed (or not) by observation or experiment. General relativity made several such predictions, including one readily testable with the technology at hand: its claim that the sun's mass bends light by an amount double that predicted by Newton's theory, 1.7 arcseconds compared to .87. Thus, once again, the decisive test of a physical claim would come at a total eclipse of the sun.

There was little chance of mounting an expedition while Europe continued to grind itself to death in the trenches. But the war couldn't last forever, and late in 1917, a handful of British scientists began to plan for the next available eclipse. It would come on May 29, 1919, following a track across the south Atlantic. In the spring of that first year of the peace, a pair of two-man teams set out, one heading to Sobral on the Brazilian mainland and the other bound for Principe, a tiny island off the West African coast.

The Principe team, astrophysicist Arthur Eddington and his assistant, reached the island on April 23. They made control images, night-sky pictures of the star field to be compared with the same stars that would be visible around the eclipsed sun. Under both Newton's and Einstein's theories, the positions of those stars would change between the control and the eclipse photographs; the question was by how much.

May 29 greeted the observers with the familiar eclipse torture: dawn came with a torrential rainstorm. The rain eased by noon, but it wasn't until 1:30, well into the partial phase, that the astronomers got their first sight of the sun. For the next several minutes clouds thickened and cleared as totality approached, and Eddington recalled that "We had to carry out our program of photographs in faith." The team took sixteen exposures, but only the final six held out much promise. Four of those six had to be developed back in England, and of the remaining two, just one had seen clear enough skies to permit preliminary analysis in the field. It took Eddington four days, but at last, on June 3, he was able to make his first comparison of the test images to the star positions recorded in the eclipsed sky.

He found the answer he'd sought: a deflection of 1.61 seconds of arc, plus or minus .3—close enough to Einstein's predicted result to claim confirmation of the general theory. He would later remember that moment as the greatest of his life. In public he was more circumspect. His telegram back to England from Principe read simply "Through cloud. Hopeful. Eddington."

Einstein himself never doubted the outcome. Two friends visited him that summer, Paul Oppenheim and his wife, Anna Oppenheim-Errara. Einstein was under the weather, so he greeted them from his bed. As they talked, a telegram arrived from Lorentz with promising news, though not final confirmation. Anna Oppenheim-Errara remembered the scene more than seventy-five years later. Einstein was in his pajamas. She could see his socks. The telegram was brought in; Einstein opened it, and said, "I knew I was right." Not, Oppenheim-Errara insisted, that he felt, or believed he was correct. "He said, 'I knew it.'"

By now, in the daily business of our warped cosmos, Vulcan barely registers, even as an antiquarian curiosity. Only a few have some vague memory of the story—mostly physicists and astronomers with a historical bent. For them, Vulcan is a cautionary tale: it's so damn easy to see what one wants or expects to find. Le Verrier himself comes off particularly poorly in these tellings, so certain of the implications of his analysis of Mercury, so eager to taste the glory of Neptune's discovery once more, that he transformed an inoffensive amateur country doctor into a *savant* in the rough. The others, like Watson, carrying to his grave the certainty that he had seen the long-sought missing planet—they can all serve as warnings that desire has no place in the rigorous, rational, implacably empirical world of science. It's such a temptation to see the past as not just past, but as less clever than the present. Perhaps that's why Vulcan's believers seem somehow comic. Like Edison with his jackrabbit, one waits for them to turn around and to find everyone rubbernecking to catch the joke.

Except—there's a story that bears on this from within Eddington's decisive photographs. The announcement of those results propelled Einstein to the worldwide fame he still enjoys, sixty years after his death. That celebrity came only after Eddington and his colleagues became convinced that they hadn't deceived themselves into finding what they had expected to see.

Remember that second team taking measurements during the eclipse at Sobral? The weather had been fine there, and the team made several more useful photographs than Eddington. When they were analyzed, they seemed to show only half the deflection claimed at Principe: Newton's answer, not Einstein's. Eddington was convinced there had to be some error with the Sobral images. But the mistake, if that's what it was, didn't show itself easily. As late as September, he wouldn't commit himself further than to say that the observed deflection lay between the two predicted values.

That was a holding action. The next month, Eddington and his colleagues confirmed that the primary Sobral instrument had an optical defect that systematically produced errors in its results. They found seven more images that had been taken with a second instrument at the Brazil site that consistently showed the Einstein value, confirming the best data from Principe. With that, Eddington felt justified in ignoring the contradicting images, and alerting the Royal Society.

Eddington was right, of course, which is all the defense one needs. The main Sobral instrument was flawed; his best images were the most nearly correct; and, of course, general relativity has survived every challenge since. It is thoroughly implicated in everything from the birth and evolution of the universe to the accuracy of the GPS system in your phone. Black holes, gravitational lensing and gravity waves, the expansion of the universe, even speculation about time travel (vanishingly unlikely, but not quite *completely* foreclosed)—all these belong to the general theory's bestiary. And more: the theory is not just richly explanatory. It has generated new ways of seeing—not just in physics, but in that broader culture, of which science is a part.

LIGHTS ALL ASKEW
IN THE HEAVENS

Men of Science More or Less Agog Over Results of Eclipse Observations.

EINSTEIN THEORY TRIUMPHS

Stars Not Where They Seemed or Were Calculated to be, but Nobody Need Worry.

The New York Times *headline on Monday, November 10, 1919.*

Just remember one human truth: the astronomers and physicists and the ordinarily curious in the mid-nineteenth century who peered longingly at the face of the sun could have said something very similar about their theory of the universe and all that it implied for the reality of Vulcan.

What moral to draw, then, of the nonexistence of an innermost planet and the universal triumph of general relativity?

At the least this: Science is unique among human ways of knowing because it is self-correcting. Every claim is provisional, which is to say each is incomplete in some small or, occasionally, truly consequential way. But in the midst of the fray, it is impossible to be sure what any gap between knowledge and nature might mean. We know *now* that Vulcan could never have existed; Einstein has shown us so. But no route to such certainty existed for Le Verrier, nor for any of his successors over the next half century. They lacked not facts, but a framework, some alternate way of seeing through which Vulcan's absence could be understood.

Such insights do not come on command. And until they do, the only way any of us can interpret what we find is through what we already know to be true. Vulcan itself, with its fifty years of stubborn possibility, is a measure of just how hard it was to see past its phantom, how genuinely powerful an achievement it was to create both Newton's account of gravity and its successor, the general theory of relativity.

Give Einstein (almost) the last word. In 1918, he spoke at the German Physical Society. There, he tried to describe what goes on inside the mind of someone attempting to interrogate nature at the edge of understanding. He didn't speak of logic, or rigor, or some exceptional mental talent. Instead, the driving force behind great work turned, he said, on "the longing to behold ... preexisting harmony." Getting there required the usual work of the researcher, of course, mathematics to learn, calculations to perform, the endless cat-and-mouse with errors of thought and execution. All that had to be done. But to do it, day after day, there was a certain way one had to be: "the state of mind which enables a man to do work of this kind" he said, "is akin to that of the religious worshiper or the lover; the daily effort does not originate from a deliberate intention or program, but straight from the heart."

For more than two centuries humankind lived in the cosmos Newton discovered. Vulcan's nonexistence did not demolish that dwelling place; rather it is the marker on which its passing is written.

Now, strange as it once seemed, beautiful as it is, Einstein's universe is our home.

Acknowledgments

First among those who have made this book into a reality: my editor, Sam Nicholson, who had actually signed me up to write something quite different. To Sam's and Random House's great credit they welcomed my shift to pursue what became this volume. Sam's editing was meticulous, firm, always kind, and measurably improved the manuscript through each draft. My agent, Eric Lupfer, skillfully guided the project through its unlikely gestation and birth, and continues to be an exemplary shepherd to my writing career. My deepest thanks to them both.

That this book made it past the "sounds interesting" phase of an idea comes down to two conversations with deeply valued friends. A couple of years ago Neil Belton, who had edited the British edition of my previous book and is publishing this one at his new home, allowed me to commandeer a dinner conversation to lay out Vulcan's strange history. He was the first to tell me that this tale should be a book. That made me think, but it took another encounter to launch the project. In the spring of 2014, Ta-Nehisi Coates talked me through the idea and after two or three afternoons he told me just to start writing, never mind what might happen with any accumulating pile of words. Without those marching orders, the work you hold would not exist. I owe Ta-Nehisi and Neil both thanks and, at the next chance, a glass or more of really good stuff.

My largest intellectual debts are to the writers and researchers who have preceded me in our collective obsession with the planet that wasn't there. You will find the works by Richard Baum and William Sheehan, N. T. Roseveare, Robert Fontenrose, and James Lequeux in the Bibliography. I've argued with interpretations found in some of those works, but as a much greater thinker once put it, I'm standing on their shoulders.

In the here and now, special thanks to my MIT colleagues David Kaiser and Allan Adams, Caltech's Sean Carroll, and Cornell's Paul Ginsparg, all of whom read the manuscript at various stages of completion—in David's case, with a gluttony for punishment sufficient to review repeated attempts to get things right. An author could not have more generous colleagues. They each made this book better; any errors that remain exist despite their best efforts and are down to me alone. Professor Matt Strassler gave me pointers to sources on the history of the problem of Mercury. In the longer term, conversations over many years with Abraham Pais, Simon Schaffer, Gerald Holton, and Peter Galison have sharpened my understanding of the history of physics and the roles played by its most notable heroes. I'm as grateful as I can be for the help of such accomplished and busy scholars.

Three cheers to all those who helped me research this book. Cara Giaimo gave valuable assistance working with newspaper archives. Archivists at the American Heritage Center and at the Hebard Historic Map Collection at the University of Wyoming were welcoming and generous with time and knowledge. The staff of the Carbon County Museum produced unexpected treasures with equal kindness. And I must also thank the Carbon County sheriff's department, who extracted my rental car from a snow-and-ice bank encountered on a too-reckless attempt to find the exact spot where Separation once stood. Yup, I was that much of a city slicker.

My gratitude as well to everyone at Random House who helped bring this book into being—with special shout-outs to associate publisher Tom Perry, copy editor Leda Scheintaub, cover designer Joseph Perez, book designer Simon M. Sullivan, and to publishing intern Lily Choi, who worked with me on picture rights and permissions.

Deep thanks to my overlapping posses of family, friends, and coworkers who nurtured this project—and me—from start to finish. My MIT colleagues were great throughout, encouraging, smart, and supportive. My thanks to Marcia Bartusiak, Alan Lightman, Seth Mnookin, and

Shannon Larkin; to my department head, Ed Schiappa, and to all my other departmental colleagues there—with an extra nod to Junot Díaz and Joe Haldeman who talked me through several of the steps of the work. John Durant, director of the MIT Museum, has been a friend and intellectual goad for years. I've benefited from the encouragement of a host of friends and colleagues from across the spectrum of science and science writing and publishing, here in no order but that of recollection: Carl Zimmer, Lisa Randall, Nikki (Veronique) Greenwood, Sean Carroll, Rose Eveleth, Neil deGrasse Tyson, Jennifer Ouellette, Brian Greene, Rebecca Saletan, David Bodanis, Ann Harris, Ed Yong, Deborah Blum, John Rubin, Ben Lillie, John Timmer, Maryn McKenna, Ian Condry, Rebecca Saxe, Ed Bertschinger, Nancy Kanwisher, Steve McCarthy, Alok Jha, Virginia Hughes, Steve Silberman, Maggie Koerth-Baker, Kevin Fong, David Dobbs, Annalee Newitz, Eric Michael Johnson, Maia Szalavitz, Tim de Chant, Tim Ferris, and Amy Harmon. A special thanks all their own goes to my students in the MIT Graduate Program in Science Writing, and especially those of the class of 2015, present at the creation: Rachel Becker, Christina Couch, Michael Greshko, Anna Nowogrodzki, Sarah Schwartz, and Josh Sokol.

My debts of gratitude to my family extend well beyond this book. In extended form, my siblings, Richard, Irene, and Leo, and spouses Jan and Rebecca; my siblings-in-law Jon, Kricket, Judy, Gay, Heinz, Neva, and Zeph; and my nieces and nephews (and my grandnephew and niece!) have sustained me with love and insight across a lifetime. I'm deeply thankful, and not only because they have mastered the art of not asking (too often) how the book is going.

Of the last I need to thank, well, words are insufficient, but they are what we have. Year in and year out, they give me patience, tolerance, laughter, distractions offered as needed, and above all love—each of these gifts and so many more. This book and its writer would not be as we are without them both. How lucky I am to owe so much to my son, Henry, and my wife, Katha.

Notes

Abbreviations

CRAS: Comptes Rendus hebdomadaires des séances de l'Académie des Sciences, available online at http://gallica.bnf.fr/ark:/12148/cb343481087/ date.langEN.

CPAE: Collected Papers of Albert Einstein, available online at http://einsteinpapers .press.princeton.edu/.

Preface

xiv "subjecting matter to number" Here I must acknowledge Alexander Koyré, who coined the notion of matter subject to number to describe Galileo's achievement. It's too wonderful a phrase to employ only once.

xiv "When he completed the calculation" Einstein to Paul Ehrenfest, *CPAE* 8, document 182, 179, and Einstein to Adriaan Fokker and to Wander Johannes de Haas, quoted in Abraham Pais, *Subtle Is the Lord*, 253.

1. "The Immovable Order of the World"

3 "That miserable business" Cook, *Edmond Halley*, 140; 148.

4 "Wren didn't believe him" Ibid., 147–48, and Westfall, *Never at Rest,* 402–3.

5 "had he died in the spring of 1684" Westfall, *Never at Rest*, 407.

6 " 'I have calculated it' " Ibid., 403.

6 "Newton went further" Ibid., 403–6.

7 "Finally, in 1687" There are many translations of *Principia* available in English. I recommend I. Bernard Cohen and Anne Whitman's: Isaac Newton, *The Principia* (Berkeley: University of California Press, 1999). It's meticulous, and the edition includes Cohen's invaluable—and book length on its own—reading guide.

9 " 'a sort of nebulous spot' " Gottfried Kirch, quoted in Gary Kronk, "From Superstition to Science," 30–35.

9 "a parabola" J. A. Ruffner, "Isaac Newton's *Historia Cometarum*," 425–51.

11 " 'The theory that corresponds exactly' " Newton, *The Principia*, trans. Cohen and Whitman, 916. Italics added.

11 " 'But we are now admitted' " Edmond Halley, "Halley's ode to Isaac Newton" *Newton, The Principia,* trans. Cohen and Whitman, 379–80.

2. "A Happy Thought"

15 "a nearly circular orbit" Baum and Sheehan, *In Search of Planet Vulcan*, 50–51.

17 "When he applied his new approach" For background on Laplace's work on Uranus, see Roger Hahn, *Pierre Simon Laplace, 1749–1827: A Determined Scientist*, 77–78, on which this account is based.

21 "to the limits of the accuracy" Gillispie et al., *Pierre-Simon Laplace 1749–1827: A Life in Exact Science*,127–28.

23 " 'when subjected to rigorous calculations' " Laplace to Le Sage, April 16, 1797, quoted in Hahn, *Pierre Simon Laplace*, 142.

23 "Every event" This argument is taken from Hahn, *Pierre Simon Laplace*; see especially page 158. The definition of determinism offered is a variation of common short-form statements of the idea; this one draws from several of them, first from the Wikipedia entry titled "Determinsm."

23 " 'failed to find his name' " This version of the contested tale comes from Roger Hahn, "Laplace and the Vanishing Role of God in the Physical Universe," in Harry Woolf, ed., *The Analytic Spirit: Essays in the History of Science* (Cornell University Press, 1981), 85–86.

24 "Herschel noted in his diary" William Herschel, quoted in Roger Hahn, *Pierre Simon Laplace*, 86.

25 " 'He merely ignores it' " Hahn in Harry Woolf, ed., *The Analytic Spirit: Essays in the History of Science*, 95.

25 " 'We may regard the present state' " Pierre Simon Laplace (trans. Truscott and Emory), *Essai philosophique sur les probabilités*, 4. Laplace proposed the idea of a comprehensively knowledgeable intellect in his 1812 *Théorie analytique des probabilités*—and had been thinking and occasionally writing about the notion at least since encountering a similar formulation in the work of his friend and colleague Condorcet in 1768.

3. "That Star Is Not on the Map"

27 "In the guidebooks" Galignai, *Galignani's New Paris Guide*, 367, and Baedeker (firm), *Paris and Environs*, 7th ed., 276.

29 " 'I must not only accept' " Lequeux, *Le Verrier—Magnificent and Detestable Astronomer*, 4.

30 " 'Laplace's inheritance' " Jean Baptiste Dumas, Sept. 25, 1877, quoted in Lequeux, *Le Verrier*, 5.

30 "It took him just two years" Le Verrier, "Sur les variations séculaires des orbites des planètes," *CRAS* 9 (1839), 370–74. Discussion in Lequeux, *Le Verrier*, 7–8, and Baum and Sheehan, *In Search of Planet Vulcan*, 70–71.

31 " 'In recent times' " Le Verrier, "Détermination nouvelle de l'orbite de Mercure et de ses perturbations," *CRAS* 16 (1843), 1054–65, quoted in Lequeux, *Le Verrier*, 13.

31 "Mercury's mass" Lequeux, *Le Verrier*, 13.

32 "He called out 'Now!' " Baum and Sheehan, *In Search of Planet Vulcan*, 73.

33 "reality and calculation diverged" Airy, "Account of Some Circumstances Historically Connected with the Discovery of the Planet Exterior to Uranus," 123.

34 "Newton's gravitational constant itself might vary with distance" Grosser, *The Discovery of Neptune*, 44; also discussed in Baum and Sheehan, *In Search of Planet Vulcan*, 80.

35 " 'perturbing Uranus' " Eugène Bouvard, "Nouvelle Table d'Uranus," 525. Cited in James Lequeux, *Le Verrier*, 24.

35 "across the channel to England" Lequeux, *Le Verrier*, 25.

35 " 'several successive revolutions' " Airy, "Account of Some Circumstances Historically Connected with the Discovery of the Planet Exterior to Uranus," 124–25.

35 "Arago pulled the younger man away" Le Verrier, "Première Mémoire sur la théorie d'Uranus," 1050, translation in Lequeux, *Le Verrier*, 26.

36 "Le Verrier instead recalculated" Grosser, *The Discovery of Neptune*, 99.

36 "some as-yet-undiscovered object" Ibid., 100.

36 "an as-yet-undiscovered trans-Uranian planet" For a contemporaneous account of this line of thinking, see John Pringle Nichol, *The Planet Neptune*, 65; 84.

38 " 'the action of a new planet' " Le Verrier, "Recherches sur les mouvements d'Uranus," 907–18, translation in Lequeux, *Le Verrier,* 28.

39 "3.3. arcseconds in diameter" Le Verrier, "Sur la planète qui produit les anomalies observées dans le mouvement d'Uranus," 428–38. To reiterate my intellectual debt: I've relied in this brief account very heavily on James Lequeux's 2013 biography of Le Verrier, and on Baum and Sheehan's *In Search of Planet Vulcan.* Like them, I found the 1962 work by Morton Grosser invaluable.

39 "But none did" Lequeux, *Le Verrier,* 33.

40 " 'a planet to discover' " Le Verrier to Galle, September 18, 1846, quoted and translated in Grosser, *The Discovery of Neptune,* 115.

Part One Interlude: "So Very Occult"

45 " 'I do not feign hypotheses' " Newton, *The Principia* (trans. Cohen and Whitman), 943.

45 " 'changeless supernatural beings' " Benson, *Cosmigraphics,*144.

47 "something Newton called a force" For an idea of just how strange "force" as a concept was to Newton's peers (and how strange in some ways it remains), see physicist Frank Wilczek's essay "Whence the Force in F-ma?"

49 " 'cannot fail but to be true' " Isaac Newton, *The Principia* (trans. Cohen and Whitman), 916.

49 "He was a secret alchemist" There is an enormous literature that has grown up around Newton's alchemical research. Betty Jo Teeter Dobbs helped found the field, and her essay "From Newton's Alchemy and His Theory of Matter" in Cohen and Westfall's *Newton: Texts, Backgrounds and Commentaries* is a fine place to start. Others to consult include Karen Figala's essay in Cohen and Smith's *The Cambridge Companion to Newton,* and the materials gathered under the direction of William Newman at the website The Cymistry of Isaac Newton (http://webapp1.dlib.indiana.edu/newton/) among many, many others.

50 "the ultimate agent" Newton in the General Scholium to *The Principia* (trans. Cohen and Whitman), 940–43.

50 " 'very little interest' " Catalogue of the Portsmouth Collection of books and papers written by or belonging to Sir Isaac Newton, xix.

4. Thirty-Eight Seconds

53 "at the tip of his pen" François Arago, quoted in James Lequeux, *Le Verrier,* 50.

53 " 'the sagacity of Le Verrier' " Ellis Loomis, *The Recent Progress of Astronomy,* 50. Emphasis in the original.

54 "addressing Uranus" See, for example, Le Verrier, U.J.J. letter to the Ministry of Public Instruction in Institut de France, *Centennaire de U.J.J. Le Verrier* (Paris: Gauthier-Villars, 1911), 50.

54 "the obvious choice" The naming controversy is widely described. This account is drawn primarily from James Lequeux, *Le Verrier,* 52–53, which quotes George Biddell Airy's letter referring to the astronomers of Northern Europe. See also Baum and Sheehan, *In Search of Planet Vulcan,* 109–10.

55 "to 'encompass in a single work' " Le Verrier to the Ministry of Public Instruction, in *Centennaire,* 51, translation in James Lequeux, *Le Verrier,* 62.

58 "yielded up Ceres and Pallas" For a discussion of Le Verrier's work on minor planets, see Lequeux, *Le Verrier,* 72–75, on which this account depends.

59 "Le Verrier's first brush with the asteroids" Le Verrier, "Sur l'influence des inclinaisons des orbites dans les perturbations des planètes," 344–48, cited in Lequeux, *Le Verrier,* 72.

59 "the same cause" Le Verrier, "Considérations sur l'ensemble du système des petites planètes situées entre Mars et Jupiter," 794.

60 "In the asteroid belt, though" Lequeux, *Le Verrier,* 74.

61 " 'a great number of other facts' " Poincaré, *The Value of Science,* 355.

62 "Urbain-Jean-Joseph Le Verrier, of course" Lequeux, *Le Verrier,* 61–65; 78–84. See also Robert Fox, *The Savant and the State,* 116–18.

62 " 'he showed little curiosity' " Joseph Bertrand, "Éloge historique de Urbain-Jean-Joseph Le Verrier," 96–97. Translation by the author.

63 " 'as his slaves' " Camille Flammarion, quoted in Lequeux, *Le Verrier,* 128.

63 "abandoned the Observatory" The "young visitor" was Camille Flammarion, who met Le Verrier in 1858, and is quoted in James Lequeux, *Le Verrier,* 128. Daverdoing's quote comes from texts gathered by a historian who organized the sale of Le Verrier's books and papers. See Lequeux, *Le Verrier,* 130. Lequeux compiled the names of those who quit the observatory between 1854 and 1867 on page 135.

64 "all behaved properly" In 1860 Le Verrier would identify a previously unde-

tected anomaly in Mars's orbit, but that problem never attracted the kind of interest or concern to come from his close examination of Mercury.

65 " 'If the tables do not strictly agree' " Le Verrier, "Nouvelles recherches sur les mouvements des planètes," 2, translated in N. T. Roseveare, *Mercury's Perihelion*, 20.

65 " 'some inaccuracy in the working' " Ibid.

65 "With a good clock" Le Verrier made this point in "Lettre de M. Le Verrier à M. Faye sur la théorie de Mercure et sur le mouvement du périhélie de cette planète," and the quality of transit information is discussed in Baum and Sheehan, *In Search of Planet Vulcan*, 135.

67 "Le Verrier's sums" Le Verrier, "Théorie et Table du mouvement de Mercure," 99.

68 "just 38 arcseconds per century" This account relies heavily on the analysis in Roseveare, *Mercury's Perihelion*, 20–24. Roseveare's book is the best technical account of the story of Mercury's precession from its first appearance as a problem to its final resolution.

5. A Disturbing Mass

69 "Nor did his fellow astronomers" He did face a public challenge in 1861, when Charles Eugène Delaunay, a long-standing antagonist, suggested that Le Verrier had simply lacked patience enough to work out some yet more precise theory of Mercury that would do away with problem. Le Verrier, rightly, dismissed the key objection to his use of observational data to correct the equations for Mercury—and he had the advantage of being right. See Lequeux, *Le Verrier*, 169.

69 " 'the mass sought' " Le Verrier, "Théorie et Table du mouvement de Mercure" (1859), 99, translation from Lequeux, *Le Verrier*, 102.

70 " 'a group of asteroids' " Le Verrier, "Lettre de M. Le Verrier à M. Faye sur la théorie de Mercure," 382.

70 " 'It's likely' " ibid.

71 " 'the area designated by M. Le Verrier' " Hervé Faye, "Remarques de M. Fay à l'occasion de la lettre de M. Le Verrier," 384.

72 " 'this huge world' " Einstein, "Autobiographical Notes," in Schilpp, ed., *Albert Einstein: Philosopher-Scientist*, 5.

73 "An object leaps into view" Fontenrose, "In Search of Vulcan," 156.

74 "He 'broke his silence,'" Le Verrier, "Remarques," 45.

74 "the two men set out" Ibid., 46.

74 "the last twelve miles" Brewster, "Romance of the New Planet," 9.

77 "discovered the first intra-Mercurian planet" Moingo's account of Le Verrier's visit, with commentary, is retold in Brewster, "Romance of the New Planet," 7–12.

77 "repeat its transits" Le Verrier reports his calculation in "Remarques," 46. It is covered, with some discussion of its implications, in Baum and Sheehan, *In Search of Planet Vulcan*, 156.

77 "very kind words for Dr. Lescarbault" The information on the response to the announcement of Lescarbault's report of this phase of the intra-Mercurian planet-finding mania comes from the account in Baum and Sheehan, *The Search for Planet Vulcan*, 155–60.

6. "The Search Will End Satisfactorily"

79 "'The singular merit'" "A supposed new interior planet," *Monthly Notices of the Royal Astronomical Society*, 2015 (1860):100.

79 "More practically" R. C. Carrington, "On some previous Observations of supposed Planetary Bodies in Transit over the Sun," 192–94.

79 "Benjamin Scott" Fontenrose, "In Search of Vulcan," 146.

79 "Rupert Wolf, a Zurich-based astronomer" Ibid., 147, Baum and Sheehan, *In Search of Planet Vulcan*, 141.

80 "Wolf's list caught the attention" J.C.R. Radeau's report was discussed in an unsigned article, "A supposed new interior planet."

80 "Radau published the results" Radau (misprinted Radan), "Future Observations of the supposed New Planet," 195–97.

80 "the planet hunt performed by multiple observers" Unsigned, "Lescarbault's Planet," 344.

81 "the 'blind economist' Henry Fawcett," "Transactions of the Sections," 142.

82 "'Mr Lummis, of Manchester'" Unsigned, "A Descriptive Account of the Planets," 129–31.

82 "'its sharp *circular* form'" Emphasis in the original.

82 "But for many others" Fontenrose, "In Search of Vulcan," 147.

83 "By the mid-1860s" Unsigned, "A Descriptive Account of the Planets," 129–32.

83 "an otherwise completely obscure M. Coumbary" Le Verrier, "Lettre de M. Le Verrier adressée à M. le Maréchal Vaillant," 1114–15

83 "Le Verrier endorsed" Ibid., 1113.

83 "a group of four eclipse mavens" E. Ledger, "Observations or supposed Observations of the Transits of Intra-Mercurial Planets," 137–38. The *"with the naked eye"* emphasis is in the original.

84 "Benjamin Apthorp Gould had a perfect Boston pedigree" This capsule biography is drawn from Trudy E. Bell's entry in the *Biographical Dictionary of Astronomers*, 833–36, Springer: 2014, online at http://link.springer.com/refe renceworkentry/10.1007%2F978-1-4419-9917-7__534.

85 "Gould sent his findings" Benjamin Gould to Yvon Villarceau, September 7, 1869, in *CRAS* T69 (1869): 813–14.

86 "Not so fast, though" Ibid., 814.

86 "He persuaded fifteen other sky-watchers" William Denning, "The Supposed New Planet Vulcan" (1869), 89.

86 "Vulcan obstinately refused" Denning reported the negative results for 1869 in *The Astronomical Register*, vol. VII, page 113. He proposed his plans and reported results for 1870 in the same journal, vol. VIII, pages 78–79 and 108–9; he announced the 1871 effort in vol. IX, page 64.

87 "Princeton's Stephen Alexander" *The New York Times* (unsigned), May 27, 1873, 4.

88 "Vulcan could be elusive," C. A. Young, "Memoir of Stephen Alexander: 1806–83," read before the National Academy, April 17, 1884, "Vulcan" online at http://www.nasonline.org/publications/biographical-memoirs/memoir -pdfs/alexander-stephen.pdf.

88 "Rupert Wolf passed word" This is a paraphrase of Wolf's letter published in *The Spectator* and reprinted in *Little's Living Age*, vol. 131, issue 1690 (November 4, 1876), 318–20.

88 "New Vulcans kept turning up" Fontenrose, "In Search of Vulcan," 149.

89 " 'Our text books on astronomy' " "The New Planet Vulcan," *Manufacturer and Builder* (unsigned), vol. 8, no. 11 (November 1876), p. 255.

89 " 'Vulcan may possibly exist,' " "Vulcan," *The New York Times* (unsigned), Sept. 26, 1876, 4.

89 " 'Vulcan exists' " Ibid.

90 "he identified five observations" Le Verrier, "Examen des observations qu'on a présentées à diverses époques comme appartenant aux passage d'une planète intra-mercurielle (suite). Discussion et conclusions." *CRAS* T83 (1876), 621–24 and 649.

90 "The headline writers would be disappointed" *Scientific American* 36, 25 (December 16, 1876), 390. The information in this section draws heavily on the sources Robert Fontenrose assembled for his "The Search for Vulcan," listed on pages 148–50.

90 "Thus Le Verrier hedged his bets" Le Verrier, "Examen des observations . . ." *CRAS* T83 (1876), 650.

90 "Le Verrier said nothing more" Baum and Sheehan, *In Search of Planet Vulcan*, 180.

91 "He did accept communion" Lequeux, *Le Verrier*, 304.

91 "The end came" Baum and Sheehan, *In Search of Planet Vulcan*, 181.

7. "So Long Eluding the Hunters"

92 " 'The country at that time' " All the details of Edison's arrival in Rawlins and the encounter at the hotel come from Edison's own account: "Edison's Autobiographical Notes," consulted at the Carbon County Museum. The passages on his Wyoming trip have been quoted elsewhere. See, e.g., Frank Lewis Dyer and Thomas Commerford Martin, *Edison: His Life and Inventions* (New York: Harper Brothers, 1929), Chapter Ten.

93 " 'not one of the "badmen" ' " University of Wyoming historian Philip Roberts notes that this incident likely didn't take place on Edison's first night in town despite Edison's own recollections to the contrary. See "Edison, The Electric Light and the Eclipse," in *Annals of Wyoming* 53, 1 (1981), 56.

94 "The federal government had funded" Baum and Sheehan, *In Search of Planet Vulcan*, 195.

95 "the element helium" Helium would be isolated on Earth a decade later by the great Scottish chemist William Ramsay.

96 " 'at the point selected' " Simon Newcomb, "Reports on the total solar eclipses on July 29, 1878, and January 11, 1880," 100.

97 "At its height" Baum and Sheehan, *In Search of Planet Vulcan*, 201.

97 "Newcomb's men found" Newcomb, "Reports on the total solar eclipses on July 29, 1878 and January 11, 1880," 100.

97 "As the days passed at Separation" Baum and Sheehan, *In Search of Planet Vulcan*, 201–2. The pattern of cloud buildup comes from Newcomb, "Reports on the total solar eclipses on July 29, 1878 and January 11, 1880," 111.

98 " 'the sun rose clear and bright' " Baum and Sheehan, *In Search of Planet Vulcan*, 202.

98 "Dust swiftly shrouded the sky" Newcomb, "Reports on the total solar eclipses on July 29, 1878, and January 11, 1880," 102.

98 "The desperation move worked" W. T. Sampson, "Reports on the total solar eclipses on July 29, 1878, and January 11, 1880," 111.

99 "playing host to two new observers" Ibid.

99 "Newcomb's own telescope" Newcomb, "Reports on the total solar eclipses on July 29, 1878, and January 11, 1880," 102.

100 "the gaps between leaves" Aristotle, the western founder of observational natural science, mentioned the effect in Book XV of *Problemmata Physica*. Properly, of course, that's the first *surviving* record of the phenomenon in the western scientific canon.

101 "his eyes adjusted to a sky grown strange" Newcomb, "Reports on the total solar eclipses on July 29, 1878, and January 11, 1880," 101; 104.

101 " 'from my previous experience' " Watson, "Reports on the total solar eclipses on July 29, 1878, and January 11, 1880," 119.

104 "he ran over to Newcomb" Ibid., 120.

104 "Newcomb later confirmed" Newcomb, "Reports on the total solar eclipses on July 29, 1878, and January 11, 1880," 105.

104 "he had no doubt about 'a' " Watson, "Reports on the total solar eclipses on July 29, 1878, and January 11, 1880," 120.

104 "as the *Laramie Weekly Sentinel* put it" *Laramie Weekley Sentinel* (unsigned), August 3, 1878, 3. Emphasis in the original.

105 "The news rocketed around the world" The Lockyer telegram and the London *Times* story both quoted in Baum and Sheehan, *In Search of Planet Vulcan*, 209–10. Swift's sighting is discussed in an unsigned item in the August 4, 1878, edition of the *New York Times*, page 1.

105 "Its first article" *New York Times*, July 30, 1878, 5.

105 "it published Watson's claim" James Watson, *New York Times*, August 8, 1878, 5.

105 " 'The negative results' " *New York Times*, August 16, 1878, 5.

106 "Watson never confessed to any doubt" James Watson, quoted in Fontenrose, "In Search of Vulcan," 153.

106 "In the beginning" Ibid., 151.

106 "He accused Watson" C. F. H. Peters quoted in an unsigned article, "The Intra-Mercurial Planet Question," 597.

107 "Watson reacted" Watson "Schreiben des Hern Prf. Watson an der Herausgeber," *Astronmische Nachtrichten* (1879) (95) 103–4, also cited in Baum and Sheehan, *In Search of Planet Vulcan*, 220–21.

107 "chided Peters for his tone" "The Intra-Mercurial Planet Question," 597–98. Italics in original.

109 "This western trip" Thomas Edison in the *Cheyenne Daily Leader*, July 19, 1878, quoted in Roberts, "Edison, The Electric Light and the Eclipse," 55.

109 "he was a tenderfoot" Edison wrote up his recollections of his western trip in autobiographical notes composed in 1908 and 1909, consulted in the Carbon County Museum, Rawlins, Wyoming. The material in this section is drawn from those and the memoir of the Separation station agent, John Jackson Clarke, "Reminiscences of Wyoming in the Seventies and Eighties," *Annals of Wyoming*, 1929, 1 and 2, 225–36. Clarke's eclipse recollections are on pages 228–29.

109 "Edison picked out a silhouette" Clarke wrote that Edison fired four times and actually hit his target each time, which, as he wrote, "imparted another angle to the joke." See Clarke, "Reminiscences of Wyoming in the Seventies and Eighties," 229.

Part Two Interlude: "A Special Way of Finding Things Out"

114 "What we see now" The difference in temperature from the hottest (densest) and coldest (most rarefied) regions of the early universe was .0005 degrees Kelvin. See http://www.astro.ucla.edu/~wright/CMB-DT.html.

115 "the smoking gun" Inflation has passed a number of observational tests, perhaps most important the match between prediction and the amount of

mass the universe contains, and another on the particular pattern of fluctuations in the CMB.

115 "It brought one of inflation's inventors to tears" See Andrea Denhoed, "Andrei Linde and the Beauty of Science," *The New Yorker*, March 18, 2014, http://www.newyorker.com/culture/culture-desk/andrei-linde-and-the-beauty-of-science.

117 "Several attempts are already under way" The POLARBEAR experiment has performed a separate measurement of B-mode polarization that supports a cosmic gravity wave interpretation, though not to the degree of confidence the BICEP team initially claimed. Other approaches include the SPIDER balloon-launched microwave telescope array and a third generation BICEP instrument that as of this writing is close to seeing first light (at microwave wavelengths, of course).

117 "ancient glow of the Big Bang" I stole this adjective from my colleague Alan Lightman, whose book on cosmology is titled *Ancient Light*. I follow the dictum "Amateurs borrow. Professionals steal."—attributed to John Lennon, who lifted it from T. S. Eliot in a recursive embodiment of the idea. (Eliot's version: "Immature poets imitate; mature poets steal...")

117 "'a special method'" Richard Feynman, *The Meaning of It All*, 5; 15.

118 "Every high school student" Online resource: https://quizlet.com/56822475/scientific-method-flash-cards/.

118 "a Coke-Mentos volcano" See, for example, science fair advice like this: http://www.sciencefairadventure.com/ProjectDetail.aspx?ProjectID=146. Further questions to be investigated: How much higher will a diet soda erupt compared to the same volume of sugared soda? What causes that difference?

118 "aimed at college students" Frank L. H. Wolfs, online resource at http://teacher.nsrl.rochester.edu/phy__labs/AppendixE/AppendixE.html. It's important to note that there is nothing exceptional about this or the science fair site above. That's the point: these are typical accounts of scientific practice, not unique ones.

119 "After July 1878" A very few continued the search for Vulcan at later eclipses, but they (still) found nothing, and whatever residual interest in that line of observation remained drained away over the next decade.

120 "fell to a variety of other objections" In 1906 another matter theory appeared similar but not identical to the zodiacal light proposal. Several re-

searchers pursued it, but it suffered from some of the same objections that skewered the earlier version. See Roseveare, *Mercury's Perihelion,* 68–94.

120 "his uncomfortably necessary conclusion" This summary of matter hypotheses is drawn from Roseveare, *Mercury's Perihelion,* 37–50.

120 "One astronomer suggested" Roseveare, *Mercury's Perihelion,* 51.

121 "the fine-structure constant" National Institute of Standards and Technology "Reference on Constants, Units and Uncertainty," online at http://physics.nist.gov/cgi-bin/cuu/Value?alph.

121 " 'All good theoretical physicists' " Richard Feynman, *QED: The Strange Theory of Light and Matter,* 129.

121 "the speed of a body might change its gravitational attraction" Roseveare, *Mercury's Perihelion,* 114–46.

8. "The Happiest Thought"

127 "Technical Examiner *Second* Class" Albrecht Fölsing, *Albert Einstein,* 231.

128 " 'the happiest thought' " From notes taken at a lecture Einstein gave in Kyoto in 1922. Cited in Abraham Pais, *Subtle Is the Lord,* 178–79.

128 "and horrible man" Lenard was the most influential founder of the anti-Semitic movement against "Jewish Physics." He joined the Nazi Party early and became a leading advocate of "German" physics under the regime.

129 "no formal training" Einstein received his doctorate for one of the papers he produced in 1905—but the process simply involved submitting the paper to the University of Zurich, where members of the physics department reviewed the work and confirmed that it met the level expected of PhD work in the field.

129 "Once Einstein represented light as quanta" Albert Einstein, "On a Heuristic Point of View Concerning the Production and Transformation of Light," *Annalen der Physik,* 17 (1905): 132–48, *CPAE* 2, document 14, 86–103.

129 "April brought Einstein's proof" Einstein, "A New Determination of Molecular Dimensions," *Annalen der Physik,* 17 (1905): 549–60, in *CPAE,* 2, document 15, 104–22. This is the paper that earned Einstein his PhD. The blue sky paper came in 1910: A. Einstein, "The Theory of the Opalescence of Homogeneous Fluids and Liquid Mixtures near the Critical State," *Annalen der Physik,* 33 (1910): 1275–98, *CPAE,* 3, 231–49, document 9.

130 "what we know as the special theory of relativity" Einstein, "On the Electrodynamics of Moving Bodies," *Annalen der Physik,* 17 (1905): 891–921, *CPAE* 2, document 23, 140–71.

130 "What does it mean, he asked" Einstein, "On the Electrodynamics of Moving Bodies," 141.

131 "If Newton were right" To be precise: the speed of light in a vacuum is 299,792,458 meters per second.

132 "no matter how precise the experiment" The best experimental test of the unchanging speed of light came with a series of experiments by the Americans Albert A. Michelson and Edward W. Morley, using exceptionally precise measuring techniques developed by Michelson. These experiments confirmed that the problem raised theoretically by Maxwell's work did in fact exist in nature, and proved to many physicists that the apparent contradiction between Maxwell and Newton had to be addressed. Einstein himself, though, either did not know of the Michelson-Morley work, or had passed over it without much attention. His impetus came almost entirely from the conflict in theory, and from other, earlier, less definitive experiments. For a detailed discussion on what Einstein knew and when he knew it, see especially Abraham Pais's *Subtle Is the Lord,* Chapter Six, which includes a technical as well as historical account. In Albrecht Fölsing's *Albert Einstein,* Chapter Nine contains a good summary of Einstein's thought processes as well. It is less detailed than Pais's, but is significantly easier to follow than Pais's more mathematical account. See also Gerald Holton, *The Thematic Origins of Scientific Thought,* Chapter 8.

132 "Einstein's insight" Einstein was also aware of a more subtle way to express a contradiction between the theory of light and Newtonian kinematics based on the problem of explaining how the laws governing electromagnetic radiation—light in all its forms, from radio waves to X-rays—could be shown to be the same whether or not they were being applied in a frame of reference at rest or one in motion. The solution, called the Lorentz transformations, after its most important author, Hendrik Lorentz, worked for Maxwell's field equations, the core body of theory about light that, among other things, posited a constant speed for light, but not for Newtonian analyses of motion.

135 "Space and time are relative" For one of the best modern popular accounts of the relativity of time and space, see Kip Thorne, *Black Holes and Time*

Warps, 71–79. Or read Einstein's own attempt to convey his theory to the public in his *Relativity: The Special and General Theory*, 21–29. The lightning and train tale, used for a long time by a lot of people, originates there.

137 "His editor had sought" Einstein, "On the Relativity Principle and the Conclusions Drawn from It," *Jarbuch der Radioaktivitat und Elektronik*, 4 (1907) 411–62, *CPAE* 2, document 47, 252–311.

138 "his happy thought" Fölsing, *Albert Einstein*, 231.

141 " 'To explain the still unexplained' " Einstein to Conrad Habicht, Dec. 24, 1907. *CPAE* 5, document 69, 47.

9. "Help Me, or Else I'll Go Crazy"

142 "Albert Einstein was unimpressed" Reported by Carl Seelig and quoted in Fölsing, *Albert Einstein*, 245.

142 " 'Gentleman, the concepts of space and time' " Hermann Minkowski, lecture given at the physics and mathematics section of the *Naturforscher* meeting held in Cologne, September 21, 1908, published as "Raum und Zeit" [Space and Time] in the *Jahresberichete der Deutschen Mathematicker-Verinigun* (1909), pp. 1–14, translated and widely quoted since. The version here follows the translation in Pais, *Subtle Is the Lord,* p. 152.

142 "with which to explore space-time" For a lovely introduction to thinking in four dimensions, take a look at Kip Thorne's modern classic *Black Holes and Time Warps,* especially Chapter Two, "The Warping of Space and Time," 87–120—which also contains a masterful exposition of the path to general relativity.

143 " 'superfluous erudition' " Fölsing, *Albert Einstein*, p. 245.

144 " 'ostentatious luxury' " Einstein to Michele Besso, May 13, 1911, *CPAE* 5, document 267, 187.

144 "For Marić, though" Dmitri Marianoff, cited in Roger Highfield and Paul Carter, *The Private Lives of Albert Einstein,* 117.

145 "His office overlooked" Frank, *Einstein: His Life and Times,* 143, quoted in Fölsing, *Albert Einstein*, 283.

145 "what gravity might do to light" The ideas described below come from this paper: A. Einstein, "On the Influence of Gravitation on the Propagation of Light," *CPAE* 3, document 23, 379–87.

149 "The tick of time runs more slowly" Feynman, *Six Not-So-Easy Pieces*, 131–36. See also Thorne, *Black Holes and Time Warps*, 102–3, for an equivalent but different statement of the same idea. Thorne's treatement is easier to follow than Feynman's, but it is a little less straightforward, as it relies on a quite subtle argument about how the two clocks are governed by two different flows of time.

149 "It bows to circumstance" The Pound-Rebka experiment, performed 1959, provided a test of the gravitational dilation of time in form directly analogous to the rocket-ship thought experiment. It was done using two sources of gamma rays (very high frequency light waves) placed on the basement and on the top floor of Harvard's Jefferson Laboratory, its physics building. As expected, the light signals from the two sources ran different rates, as calculated from Einstein's theory.

151 "He told one friend" Einstein to Zannger, undated, probably June 1912, *CPAE* 5, document 406, 307; Einstein to M. Besso, *CPAE* 5, document 377, 276.

151 "'the foundations of geometry'" Einstein, lecturing in Kyoto in 1922, quoted in Pais, *Subtle Is the Lord*, 212.

152 "Einstein begged" Ibid.

152 "no one has found an error" Robert Osserman, *The Poetry of the Universe*, 5.

155 "'It is all going marvelously'" Einstein to Ludwig Hopf, Aug. 16, 1912, *CPAE* 5, document 416, 321.

156 "the problem of Mercury" Einstein and Michele Besso, "Manuscript on the Motion of the Perihelion of Mercury," 1913, *CPAE* 4, document 14, 360–473 (German original).

156 "some guilty pleasures" The calculation error is described in Michel Janssen, "The Einstein-Besso Manuscript: Looking over Einstein's Shoulder," 9, online at http://zope.mpiwg-berlin.mpg.de/living__einstein/teaching /1905__S03/pdf-files/EBms.pdf.

  The larger error can be found in A. Einstein and M. Besso (1913), *CPAE* 4, document 14, 444 (page 41 of the original, reproduced in facsimile on page 670), discussed by Janssen on page 14 of his paper.

156 "a rare window into the act of scientific thinking" I'm indebted to Janssen—an editor of the Einstein Papers—for this discussion of the Einstein-

Besso collaboration. My account of what it reveals about how Einstein thought derives from his work, most notably his "The Einstein-Besso Manuscript: Looking over Einstein's Shoulder."

158 "Einstein never published" In 1914, another physicist, Johannes Droste, worked out the same answer and did publish that work to no apparent impact on the larger question of the validity of general relativity. Janssen, "The Einstein-Besso Manuscript: Looking over Einstein's Shoulder," 12.

159 " 'the merely personal' " Einstein, "Autobiographical Note," in Schilpp, *Albert Einstein: Philosopher-Scientist*, 5.

10. "Beside Himself with Joy"

160 " 'The Berlin public' " Theodor Wolff, in *Das Vorspiel*, vol. 1, 1924, quoted in Dieter and Ruth Glatzer, *Berliner Leben*, 506.

162 " 'That a man can take pleasure' " Einstein, "The World as I See It," originally published in 1930; reprinted in Einstein, *Ideas and Opinions*, 10.

162 "As dusk approached" The account of this first gas attack is drawn from Martin Gilbert, *The First World War*, 144–45.

163 "as General Sir John French reported" Gilbert, *The First World War*, 144.

163 " 'Our whole, highly praised technological progress' " Einstein to Heinrich Zannger, December 6, 1917, *CPAE* 8, document 403, 411–12.

164 " 'this huge world,' " Einstein, "Autobiographical Notes" in Schilpp, *Albert Einstein: Philosopher-Scientist*, 5.

165 "it became possible to contrast competing ideas" I am indebted to Albrecht Fölsing for his account of these lectures in his *Albert Einstein*, 357–59.

165 "Einstein told them to their faces" Einstein, "The Formal Foundation of the General Theory of Relativity," Proceedings of the Prussian Academy of Sciences, II (1914): 1030–85. In *CPAE* 6, document 9, 30–85.

165 "Einstein did receive a few letters" Fölsing, *Albert Einstein*, 359. For what correspondence Einstein did receive, see Hendrik A. Lorentz to Einstein, between Jan. 1 and 23, 1915, and Tullia Levi-Civita to Einstien, March 28, 1915, in *CPAE* 8, documents 43 and 67, 49–56; 79–80.

165 " 'no one will believe you' " Einstein quoted in Miller, *Einstein, Picasso: Space, Time and the Beauty That Causes Havoc*, 228.

166 "violated a key claim of special relativity" More technically: his 1913–14 theory violated the invariance of physical law under transformation between two reference frames in relative motion.

166 "the beautiful and melancholy Christmas Truce" There are many accounts of the cease-fires informally agreed on Christmas Day 1914. For one very well written account of many, see Modris Eksteins, *Rites of Spring*, 95–98.

166 "an utterly unsuccessful jaunt" Fölsing, *Albert Einstein*, 360–63.

167 "left Hilbert ready to accept" Einstein to Wander and Geertruida de Haas, August 2, 1915, *CPAE* 8, document 144, 116–17.

167 "Hilbert believed him" Hilbert ultimately produced his version of general relativity a few days before Einstein produced his ultimate version of the theory. There was a moment of coolness between the two men in December 1915, as Einstein believed Hilbert might have been trying to take some share of the credit for his discovery, but in rapid order Hilbert made it clear that he was not contesting priority, and they swiftly resumed cordial relations. The prevailing view has been that the two men came up independently and essentially simultaneously with the same, correct answer. But the discovery of an archived set of Hilbert's proofs and a close analysis by three historians of physics has shown that Hilbert's November version was, in fact, incomplete, and that Hilbert revised what he later published in light of Einstein's final conclusions. See Leo Corry, Jürgen Renn, and John Stachel, "Belated Decisions in the Hilbert-Einstein Priority Dispute," *Science* 278, November 14, 1997.

167 " 'a blatant contradiction' " Einstein to Erwin Freundlich, *CPAE* 8, document 123, 132–33.

168 "all his time to thought and calculation" Einstein to Arnold Sommerfeld, November 28, 1915, *CPAE* 8, document 153, 152–153.

169 "the first of four updates" Einstein, "On the General Theory of Relativity," *CPAE* 6, document 21, 98–107.

170 "The next Thursday" Einstein, "On the General Theory of Relativity (Addendum)" *CPAE* 6, document 22, 108–10.

171 " 'The calculation for the planet Mercury' " Einstein, "Explanation of the Perihelion Motion of Mercury from the General Theory of Relativity," *CPAE* 6, document 24, 112–16.

173 "genuine palpitations" Einstein to Adriaan Fokker, quoted in Pais, *Subtle Is the Lord*, 253.

173 " 'beside himself with joy' " Einstein to Paul Ehrenfest, January 17, 1916, *CPAE* 8, document 182, 179.

173 " 'The years of searching in the dark' " Einstein, *The Origins of the General Theory of Relativity* (Glasgow: Jackson, Wylie, 1933), quoted in Pais, *Subtle Is the Lord*, 257.

Postscript: "The Longing to Behold . . . Preexisting Harmony"

174 "his final theory of gravity" Einstein, "The Field Equations of Gravitation," *CPAE* 6, document 25, 117–20.

174 " 'a little worn out' " Einstein to Michele Besso, December 10, 1915, *CPAE* 8, document 162, 159–60.

174 "Study the equations well" Einstein to Arnold Sommerfeld, December 9, 1915, *CPAE* 8, document 161, 159.

175 "how matter and energy together tell space-time" The physicist John Wheeler first popularized this framing for general relativity.

176 " 'I have forgotten how to hate' " Einstein to Besso, May 13, 1917, *CPAE* 8, document 339, 329–30.

176 "plan for the next available eclipse" Matthew Stanley, "An Expedition to Heal the Wounds of War: The 1919 Eclipse and Eddington as Quaker Adventurer," *Isis* 94, 1 (2003): 72.

177 " 'carry out our program of photographs in faith' " Stanley, "An Expedition to Heal the Wounds of War," 76. This author had a similar experience at the eclipse of 1991, while making a film for the *NOVA* series on PBS. Clouds covered the sun (and my cameras) at the summit of Mauna Kea in Hawaii fifteen minutes before totality, and I found myself bargaining with all manner of notions of the divine in the hopes that the sky would clear—which it did.

177 " 'Through cloud. Hopeful' " Ibid.

178 " ' "I knew it" ' " Anna Oppenheim-Errara, personal communication in 1995. Anna Oppenheim met Einstein in 1911 at the first Solvay conference. She was a teenager, engaged to be married. Her father was provost at the University of Brussels, and her fiancé was a physicist. At a reception her fa-

ther threw for the distinguished scientific visitors, her husband-to-be pointed out the rather scruffy and much younger than the average Einstein and told her to bring him an extra sandwich, for, he told her, despite appearances, he was the best of the lot.

179 "Eddington felt justified" Stanley, "An Expedition to Heal the Wounds of War," 78.

181 " 'the state of mind which enables a man' " Einstein, speaking at the German Physical Society in 1918, quoted in Pais, *Subtle Is the Lord*, 26–27.

Bibliography

Any work of historical interpretation both depends on and argues with its predecessors; this book is no different. A conventional list of the books and articles cited in this book follows below, but there are a few writers I want to both acknowledge and draw special attention to as the ones I most valued, and in some cases, as those with whom this work contends.

First among not-necessarily equals: for anyone interested in Isaac Newton, there is no substitute for Richard Westfall's *Never At Rest*. It is the definitive biography. It will offer the technically inclined a solid grounding in Newton's science. It also delivers a comprehensive and wonderfully readable narrative life. Its bibliography and apparatus offer an invitation to follow any aspect of Newton's career to whatever depth a reader may desire. I. B. Cohen and Ann Whitman's translation of the *Principia* is the one to get. It is among the best designed of the editions available, which is important for a book that relies as heavily as it does on diagrams, but the real value of this edition comes from Cohen's reader's guide—a book of its own, more than three hundred pages of explanation and interpretation. Accept no substitutes.

The story of the nineteenth-century planet hunters is one that has attracted a wide range of professional and a smaller cadre of popular writers. Several works were invaluable in the construction of this book, both as narrative and exposition in their own right and as guides to the underlying primary sources. For the details of Le Verrier's progress I depended most on James Lequeux's recent, somewhat technical biography of Le Verrier. For Vulcan, both its backstory in the discovery of Neptune and its post-1859 fate, I am deeply indebted to Richard Baum and William Sheehan's *In Search of Planet Vulcan*. Among that book's many virtues is its meticulous sourcing, and I found it to be invaluable both for its clear narrative and as a gateway to the primary literature of nineteenth-century astronomy. This was in fact the book that in some sense made me want to write this one, for while it's a meticulously researched account, I found myself regularly arguing with its interpretations . . . which disputations are embedded here.

Both Baum and Sheehan and I drew heavily on the work of the historian Robert Fontenrose, whose paper on the claims of Vulcan sightings is the comprehensive guide to the professional and popular accounts of the pursuit of the hypothetical planet in the second half of the nineteenth century. Finally, N. T. Roseveare's *Mercury's Perihelion* is a meticulous technical account of the different explanations proposed for Mercury's orbit from the discovery of the excess perihelion advance to Einstein's ultimate solution—and beyond, to proposed and (so far) unsuccessful attempts to construct alternatives to general relativity with which nature agrees.

On Albert Einstein and the path to general relativity, I owe many debts—please see the acknowledgments for those who shared their time with me over many years' obsession with this extraordinary figure. In preparing this account, three books were especially helpful. The first is what remains, after more than three decades, the best one-volume technically literate biography of Einstein, Abraham Pais's *Subtle Is the Lord.* Subsequent research by scholars working through the Einstein papers has turned up a significant amount of new information on the precise development of Einstein's thinking on a number of topics, but Pais's work remains the essential starting point for any comprehensive consideration of his friend's full range of scientific inquiry and achievement. Albrecht Fölsing's *Albert Einstein* is an exemplary account of the life, and it offers a version of the scientific journey that is much more accessible than Pais's. Similar in scope, Walter Isaacson's *Einstein* is both the most up to date and the most fun to read of the major popular biographies; if you're not looking for a mathematical introduction (go to Pais for that), this is where you start an Einstein journey.

Finally, there are two more books that I relied on very heavily, both mine. The research that enabled me to write *Newton and the Counterfeiter* and *Einstein in Berlin* came into play here, and as noted earlier, passages in Chapter One, Part Three, and the Postscript first appeared in different forms in those works.

Airy, George Biddell. "Account of Some Circumstances Historically Connected with the Discovery of the Planet Exterior to Uranus." *Monthly Notices of the Royal Astronomical Society* 7 (November 8, 1846).

Anonymous (leader). "Miscellaneous Intelligence: A Supposed New Interior Planet." *Monthly Notices of the Royal Astronomical Society* 20, 3 (January 13, 1860): 98–100.

Baedeker, Karl (firm). *Paris and Environs with routes from London to Paris and from Paris to the Rhine and Switzerland: Handbook for travellers.* 7th ed. Remodeled and augm. Leipsig: K. Baedeker, 1881.

Baum, Richard L., and William Sheehan. *In Search of Planet Vulcan.* New York: Plenum Press, 1997.

Bell, Trudy E. "Gould, Benjamin Apthorp." Entry in the *Biographical Dictionary of Astronomers.* New York: Springer, 2014, 833–36.

Benson, Michael. *Cosmigraphics.* New York: Abrams, 2014.

Bertrand, M. J. "Éloge historique de Urbain-Jean-Joseph Le Verrier." *Annales de l'Observatoire de Paris* 15 (1880): 3–22, http://www.academie-sciences.fr/activite/archive/dossiers/eloges/leverrier__vol3255.pdf: 81–114. Pagination in endnotes from the web edition.

Bouvard, Eugène. "Nouvelle Table d'Uranus." *CRAS* 21 (1845): 524–25.

Brewster, David. "Romance of the New Planet." *North British Review,* Edinburgh, T. and T. Clark 33 (August–November 1860): 1–21.

British Association. *Report of the Thirty First Meeting of the British Association for the Advancement of Science; Held at Manchester in September 1861.* London: John Murray, 1862.

Browne, Janet. *Charles Darwin: The Power of Place* (vol. II of a biography). Princeton: Princeton University Press, 2002.

Carrington, R. C. In the 10th number of Professor Wolf's *Mittheilungen über die Sonnenflecken,* several cases are quoted of the observation of planetary bodies in transit over the sun, "some of which are evidently of another character, but the following deserving of attention." *Monthly Notices of the Royal Astronomical Society* 20, 3 (January 13, 1860): 100–101.

———. "On some previous Observations of supposed Planetary Bodies in Transit over the Sun." *Monthly Notices of the Royal Astronomical Society* 20, 5 (March 9, 1860): 192–94.

Clarke, John Joseph. "Reminiscences of Wyoming in the Seventies and Eighties." *Annals of Wyoming* 1 and 2 (1929): 225–36.

Cohen, I. Bernard, and George E. Smith, eds. *The Cambridge Companion to Newton.* Cambridge: Cambridge University Press, 2002.

Cohen, I. Bernard, and Richard S. Westfall. *Newton: Texts, Backgrounds and Commentaries.* New York: W. W. Norton, 1995.

Cook, Alan H. *Edmond Halley: Charting the Heavens and the Seas.* Oxford: Oxford University Press, 1998.

Corry, Leo, Jürgen Renn, and John Stachel. "Belated Decisions in the Hilbert-Einstein Priority Dispute." *Science* 278 (November 14, 1997): 1270–73.

Coumbary, Aristide. "Lettre de M. Aristide Coumbary." *CRAS* T60 (1865): 1114–15.

Denning, William. "The Supposed New Planet Vulcan." *The Astronomical Register* VII (1869): 89.

———. "The Supposed Planet Vulcan." *The Astronomical Register* VIII (1870): 77–78, 108–9.

———. "The Supposed Planet Vulcan." *The Astronomical Register* IX (1871): 64.

Dobbs, B. J.T. *The Janus Faces of Genius: The Role of Alchemy in Newton's Thought.* Cambridge: Cambridge University Press, 1991.

Edison, Thomas. "Autobiographical Notes." Accessed at the Carbon County Museum, Rawlins, Wyoming, on January 23, 2015.

Einstein, Albert. *The Collected Papers of Albert Einstein*, online at http://einstein papers.press.princeton.edu/.

———. *Relativity: The Special and General Theory.* New York: Dover, 15th edition, 1952 (the first edition was published in 1916).

———. *Ideas and Opinions.* New York: Crown Publishers, 1954.

Eksteins, Modris. *Rites of Spring.* Boston: Houghton Mifflin/Mariner, 2000.

Fawcett, Henry. "Transactions of the Sections." *Report of Thirty First Meeting of the British Association for the Advancement of Science.* London: John Murray, 1862.

Faye, Hervé. "Remarques de M. Fay à l'occasion de la lettre de M. Le Verrier." *CRAS*, T49 (1859): 383–85.

Feynman, Richard. *The Characteristic of Physical Law*. London: BBC, 1965.

———. *The Meaning of It All*. New York: Perseus Books, 1998.

———. *Six Not-So-Easy Pieces*. New York: Basic Books, 1997.

———. *QED: The Strange Theory of Light and Matter*. Princeton: Princeton University Press, 1985.

Fölsing, Albrecht. *Albert Einstein*. New York: Viking Penguin, 1997.

Fontenrose, Robert. "In Search of Vulcan." *The Journal for the History of Astronomy* 4 (1973): 145–58.

Fox, Robert. *The Savant and the State: Science and Cultural Politics in Nineteenth-Century France*. Baltimore: Johns Hopkins University Press, 2012.

Frank, Philipp. *Einstein: His Life and Times*. New York: Alfred A. Knopf, Inc., 1947, rev. 1953.

Galignani A. and W. *Galignani's New Paris Guide*. Paris: A. and W. Galignani, 1852.

Galison, Peter. *Einstein's Clocks and Poincaré's Maps*. New York: W. W. Norton & Co., 2003.

Gilbert, Martin. *The First World War*. New York: Henry Holt, 1994.

Gillispie, Charles Coulston, with the collaboration of Robert Fox and Ivor Grattan-Guinness. *Pierre-Simon Laplace 1749–1827: A Life in Exact Science*. Princeton: Princeton University Press, 1997.

Glatzer, Dieter, and Ruth Glatzer. *Berliner Leben*, 2 vols. Berlin: Rütten & Verlag, 1988.

Goodstein, David L., and Judith R. Goodstein. *Feynman's Lost Lecture*. New York: W. W. Norton & Company, 1996.

Gould, Benjamin. "Sur l'éclipse solaire du 7 août dernier." *CRAS* 69 (1869): 813–14.

Grosser, Morton. *The Discovery of Neptune*. Cambridge, Massachusetts: Harvard University Press, 1962.

Hacking, Ian. *The Emergence of Probability: A Philosophical Study of the Early Ideas About Probability, Induction and Statistical Inference*. Cambridge: Cambridge University Press, 1975.

Hahn, Roger. *Pierre Simon Laplace, 1749–1827: A Determined Scientist*. Cambridge, Massachusetts: Harvard University Press, 2005.

Highfield, Roger, and Paul Carter. *The Private Lives of Albert Einstein*. New York: St. Martin's Griffin, 1994.

Holton, Gerald. "Einstein's Third Paradise." *Daedalus* (Fall 2003): 26–34.

———. *The Thematic Origins of Scientific Thought: Kepler to Einstein*. Cambridge, Massachusetts: Harvard University Press, 1988.

Institut de France. *Centennaire de U. J. J. Le Verrier*. Paris: Gauthier-Villars, 1911.

Isaacson, Walter. *Einstein: His Life and Universe*. New York: Simon and Schuster, 2007.

Janiak, Andrew. "Newton's Philosophy." *Stanford Encyclopedia of Philosophy* (Summer 2014), Edward N. Zalta (ed.), http://plato.stanford.edu/archives/sum2014/entries/newton-philosophy/.

Janssen, Michel. "The Einstein-Besso Manuscript: Looking Over Einstein's Shoulder," http://zope.mpiwg-berlin.mpg.de/living__einstein/teaching/1905__S03/pdf-files/EBms.pdf.

———. "The twins and the bucket: How Einstein made gravity rather than motion relative in general relativity." *Studies in History and Philosophy of Modern Physics* 43 (2012): 159–75.

Kronk, Gary. "From Superstition to Science." *Astronomy* 41, 11 (November 2013): 30–35.

Kühn, Sebastian, and Bill Rebiger. "Hidden Secrets or the Mysteries of Daily Life. Hebrew Entries in the Journal Books of the Early Modern Astronomer Gottfried Kirch." *European Journal of Jewish Studies* 6, 1 (2012): 149–50.

Laplace, Pierre-Simon. *Essai philosophique sur les probabilités*. Translated by Frederick Wilson Truscott and Frederick Lincoln Emory. New York: John Wiley & Sons, 1940.

———. *Mechanism of the Heavens*. Translated by Mary Somerville. Cambridge: Cambridge University Press, 1831 and 2009.

Ledger, E. "Observations or supposed observations of the transits of intra-Mercurial planets or other bodies across the sun's disk." *The Observatory* 3, 29 (1879): 135–38.

Lequeux, James. *Le Verrier—Magnificent and Detestable Astronomer*. New York: Springer, 2013.

Levenson, Thomas. *Einstein in Berlin*. New York: Bantam, 2003.

———. *Newton and the Counterfeiter*. New York: Houghton Mifflin Harcourt, 2009.

Leverington, David. *Babylon to Voyager and Beyond: A History of Planetary Discovery*. Cambridge: Cambridge University Press, 2003.

Le Verrier, Urbain-Jean-Joseph. "Examen des observations qu'on a présentées à diverses époques comme appartenant aux passage d'une planète intra-mercurielle (suite). Discussion et conclusions." *Comptes Rendus* 83 (1876): 621–23.

———. "Consdérations sur l'ensemble du système des petites planètes situées entre Mars et Jupiter." *CRAS* T37 (1853): 793–98.

———. "Détermination nouvellé dé l'orbite de Mercure et de ses perturbations." *CRAS* 16 (1843): 1054–65.

———. "Les planètes intra-mercurielles (suite)." *CRAS* 83 (1876): 647–50.

———. "Lettre de M. Le Verrier á M. Faye sur la théorie de Mercure et sur le movement du périhélie de cette planète." *CRAS* 49 (1859): 379–83.

———. "Lettre de M. Le Verrier adressée à M. le Maréchal Vaillant" and "Lettre de M. Aristide Combary." *CRAS* 60 (1865): 1113–15.

———. "Nouvelles recherches sur les mouvements des planètes." *CRAS* 29 (1849): 1–5.

———. "Première Mémoire sur la théorie d'Uranus." *CRAS* 21 (1845): 1050–55.

———. "Recherches sur les mouvements d'Uranus." *CRAS* 22 (1846): 907–18.

———. "Remarques" [on M. Lescarbault's observation of a planet inside the orbit of Mercury]. *CRAS* 50 (1860): 45–46.

———. "Sur la planète qui produit les anomalies observées dans le mouvement d'Uranus.—Détermination de sa masse, de son orbite et de sa position actuelle." *CRAS* 23 (1846): 428–38.

———. "Sur les variations séculaires des orbites des planètes." *CRAS* 9 (1839): 370–74.

———. "Sur l'influence des inclinaisons des orbites dans le perturbations des planètes. Détermination d'une grande inégalité du moyen mouvement de Pallas." *CRAS* 13 (1841): 344–48.

———. "Théorie et Table du mouvement de Mercure." *Annales de l'Observatoire Impérial de Paris* 5 (1859): Chapter XV, 1–196.

Loomis, Elias. *The Recent Progress of Astronomy; especially in the United States.* New York: Harper & Brothers, 1850. (Google ebook: https://play.google.com/store/books/details?id=00IDAAAAQAAJ&rdid=book-00IDAAAAQAAJ&rdot=1).

McMullin, Ernan. "The Impact of Newton's *Principia* on the Philosophy of Science." *Philosophy of Science* 68, 3 (September 2001): 279–310.

Meeus, J. "The maximum possible duration of a total solar eclipse." *Journal of the British Astronomical Association* 113, 6 (December 2003): 343–48.

Miller, Arthur. *Einstein, Picasso: Space, Time and the Beauty That Causes Havoc.* New York: Basic Books, 2002.

New York Times. "Vulcan." May 27, 1873, p. 4.

New York Times. "Vulcan." September 26, 1876, p. 4.

Newton, Isaac. *The Principia: Mathematical Principles of Natural Philosophy.* Translated by I. Bernard Cohen and Anne Whitman. Berkeley: University of California Press, 1999.

The Newton Papers Project, online at http://www.newtonproject.sussex.ac.uk /prism.php?id=1.

Nichol, John Pringle. *The Planet Neptune: An Exposition and History.* Edinburgh: John Johnstone, 1848. (Google ebook: http://books.google.com/books?id =BxUEAAAAQAAJ&pg=PP1&lpg=PP1&dq=the+planet+neptune+pringle +nichol&source=bl&ots=S3VK9uuICa&sig=oR8tgZVNb14X6-PI5ibQ4oa CGsQ&hl=en&sa=X&ei=y6h0VLeeBqq__sQS3zoKwAw&ved=oCE4Q6A EwBQ#v=onepage&q=the%20planet%20neptune%20pringle%20nichol &f=false).

Osserman, Robert. *The Poetry of the Universe.* New York: Anchor, 1995.

Pais, Abraham. *Subtle Is the Lord.* New York: Oxford University Press, 1982. (The one essential biography of Einstein, written by a wonderful man who is mentioned in the acknowledgments.)

Peters, C.F.H. (Quoted in) [Unsigned] "The Intra-Mercurial Planet Question." *Nature.* 20, 521(1879): 597–99.

Poincaré, Henri. *Science and Hypothesis.* New York: Dover Publications, 1952.

———. *Science and Method.* New York: Dover Publications, 1952.

———. *The Value of Science.* New York: Dover Publications, 1958.

Proctor, R. A. "New Planets Near the Sun." London: Strahan and Company, *The Contemporary Review* XXXIV (March 1879): 660–77.

Radau (misprinted Radan), J.C.R."Future Observations of the supposed New Planet." *Monthly Notices of the Royal Astronomical Society* 20, 5 (March 9, 1860): 195–97.

Roberts, Philip. "Edison, The Electric Light and the Eclipse." *Annals of Wyoming* 53, 1 (1981): 54–62.

Roseveare, N. T. *Mercury's Perihelion: From Le Verrier to Einstein*. Oxford: Clarendon Press, 1982.

Royal Astronomical Society (unsigned). "A supposed new interior planet." *Monthly Notices of the Royal Astronomical Society* 20, 5 (1860): 98–100.

———. "Lescarbault's Planet." *Monthly Notices of the Royal Astronomical Society* 20, 8 (1860): 344.

Ruffner, J. A. "Isaac Newton's *Historia Cometarum* and the Quest for Elliptical Orbits." *Journal for the History of Astronomy* 41, 145, part 4 (November 2010): 425–51.

Schaffer, Simon. "Newtonian Angels," in Joad Raymond, ed. *Conversations with Angels: Essays Towards a History of Spiritual Communication, 1100–1700*. Basingstoke: Palgrave Macmillan, 2011.

Schilpp, Paul Arthur, ed. *Albert Einstein: Philosopher-Scientist*. La Salle, Illinois: Open Court, 1949. Third Edition, 1982.

Schlör, Joachim. *Nights in the Big City: Paris, Berlin, London 1840–1930*. London: Reaktion Books, 1998.

Seelig, Carl. *Albert Einstein: A Documentary Biography*. London: Staples Press, 1956.

Stanley, Matthew. "An Expedition to Heal the Wounds of War: The 1919 Eclipse and Eddington as Quaker Adventurer." *Isis* 94, 1 (March 2003): 57–89.

Thorne, Kip. *Black Holes and Time Warps: Einstein's Outrageous Legacy*. New York: Norton, 1995.

United States Naval Observatory. *Washington Observations*, 1876 and 1880.

Unsigned. "A Descriptive Account of the Planets." *The Astronomical Register*, IV, 41 (1866): 129–32.

Unsigned, "A supposed new interior planet." *Monthly Notices of the Royal Astronomical Society* 20, 5 (1860): 98–101.

Unsigned. "The Intra-Mercurial Planet Question." *Nature* 20, 521 (1879): 597–99.

Unsigned. "Lescarbault's Planet." *Monthly Notices of the Royal Astronomical Society* 20, 8 (June 8, 1860): 344.

Unsigned. "The Planet Vulcan." *Littell's Living Age* 131, 1690 (1876): 318–20.

Unsigned "The New Planet Vulcan." *Manufacturer and Builder.*" 8, 11 (November 1876): 255

Various, including Simon Newcomb, W. T. Sampson, and James C. Watson. "Reports on the total solar eclipses on July 29, 1878 and January 11, 1880." *Washington Observations 1876*, Appendix III, Washington: United States Naval Observatory, 1880.

Walker, Sears C. "Researches Relative to the Planet Neptune." In *Smithsonian Contributions to Knowledge*, Vol. II, Washington, D.C.: Smithsonian Institution, 1851.

Watson, James C. "Schreiben des Herrn Prof. Watson an den Herausgeber." *Astronmische Nachtrichten* 95(1879): 101–6.

Webb, T. W. *Celestial Objects for Common Telescopes*. London: Longman, Green, Longman, and Roberts, 1859.

Westfall, Richard. *Never at Rest*. Cambridge: Cambridge University Press, 1983.

Wilczek, Frank. "Whence the Force in F=ma?" *Physics Today* (2004), retrieved at http://ctpweb.lns.mit.edu/physics__today/phystoday/%20Whence__cshock.pdf.

Wilson, Curtis. "The Great Inequality of Jupiter and Saturn: From Kepler to Laplace." *Archive for History of Exact Sciences* 33 (1985): 15–290.

Illustration Credits

Index

[Page numbers in *italic* refer to captions.]

C

D

E

F